A Member of the International Code Family™

ICC
INTERNATIONAL
CODE COUNCIL®

INTERNATIONAL PLUMBING CODE®

2003

2003 International Plumbing Code®

First Printing: February 2003
Second Printing: May 2003

ISBN # 1-892395-62-2 (soft-cover edition)
ISBN # 1-892395-61-4 (loose-leaf edition)
ISBN # 1-892395-84-3 (e-document)

PRINTED IN THE U.S.A.

PREFACE

Introduction

Internationally, code officials recognize the need for a modern, up-to-date plumbing code addressing the design and installation of plumbing systems through requirements emphasizing performance. The *International Plumbing Code®*, in this 2003 edition, is designed to meet these needs through model code regulations that safeguard the public health and safety in all communities, large and small.

This comprehensive plumbing code establishes minimum regulations for plumbing systems using prescriptive and performance-related provisions. It is founded on broad-based principles that make possible the use of new materials and new plumbing designs. This 2003 edition is fully compatible with all the *International Codes* ("I-Codes") published by the International Code Council (ICC), including the *International Building Code, ICC Electrical Code, International Energy Conservation Code, International Existing Building Code, International Fire Code, International Fuel Gas Code, International Mechanical Code, ICC Performance Code, International Private Sewage Disposal Code, International Property Maintenance Code, International Residential Code, International Urban-Wildland Interface Code* and *International Zoning Code.*

The *International Plumbing Code* provisions provide many benefits, among which is the model code development process that offers an international forum for plumbing professionals to discuss performance and prescriptive code requirements. This forum provides an excellent arena to debate proposed revisions. This model code also encourages international consistency in the application of provisions.

Development

The first edition of the *International Plumbing Code* (1995) was the culmination of an effort initiated in 1994 by a development committee appointed by the ICC and consisting of representatives of the three statutory members of the International Code Council: Building Officials and Code Administrators International, Inc. (BOCA), International Conference of Building Officials (ICBO) and Southern Building Code Congress International (SBCCI). The intent was to draft a comprehensive set of regulations for plumbing systems consistent with and inclusive of the scope of the existing model codes. Technical content of the latest model codes promulgated by BOCA, ICBO and SBCCI was utilized as the basis for the development. This 2003 edition presents the code as originally issued, with changes approved through the ICC Code Development Process through 2002. A new edition such as this is promulgated every three years.

With the development and publication of the family of *International Codes* in 2000, the continued development and maintenance of the model codes individually promulgated by BOCA ("BOCA National Codes"), ICBO ("Uniform Codes") and SBCCI ("Standard Codes") was discontinued. This 2003 *International Plumbing Code*, as well as its predecessor—the 2000 edition—is intended to be the successor plumbing code to those codes previously developed by BOCA, ICBO and SBCCI.

The development of a single set of comprehensive and coordinated *International Codes* was a significant milestone in the development of regulations for the built environment. The timing of this publication mirrors a milestone in the change in structure of the model codes, namely, the pending Consolidation of BOCA, ICBO and SBCCI into the ICC. The activities and services previously provided by the individual model code organizations will be the responsibility of the Consolidated ICC.

This code is founded on principles intended to establish provisions consistent with the scope of a plumbing code that adequately protects public health, safety and welfare; provisions that do not unnecessarily increase construction costs; provisions that do not restrict the use of new materials, products or methods of construction; and provisions that do not give preferential treatment to particular types or classes of materials, products or methods of construction.

Adoption

The International Plumbing Code is available for adoption and use by jurisdictions internationally. Its use within a governmental jurisdiction is intended to be accomplished through adoption by reference in accordance with proceedings establishing the jurisdiction's laws. At the time of adoption, jurisdictions should insert the appropriate information in provisions requiring specific local information, such as the name of the adopting jurisdiction. These locations are shown in bracketed words in small capital letters in the code and in the sample ordinance. The sample adoption ordinance on page v addresses several key elements of a code adoption ordinance, including the information required for insertion into the code text.

Maintenance

The *International Plumbing Code* is kept up to date through the review of proposed changes submitted by code enforcing officials, industry representatives, design professionals and other interested parties. Proposed changes are carefully considered through an open code development process in which all interested and affected parties may participate.

The contents of this work are subject to change both through the Code Development Cycles and the governmental body that enacts the code into law. For more information regarding the code development process, contact the Code and Standard Development Department of the International Code Council.

While the development procedure of the *International Plumbing Code* assures the highest degree of care, ICC and the founding members of ICC—BOCA, ICBO, SBCCI—their members and those participating in the development of this code do not accept any liability resulting from compliance or noncompliance with the provisions because ICC and its founding members do not have the power or authority to police or enforce compliance with the contents of this code. Only the governmental body that enacts the code into law has such authority.

Letter Designations in Front of Section Numbers

In each code development cycle, proposed changes to this code are considered at the Code Development Hearing by the International Plumbing Code Development Committee, whose action constitutes a recommendation to the voting membership for final action on the proposed change. Proposed changes to a code section whose number begins with a letter in brackets are considered by a different code development committee. For instance, proposed changes to code sections which have the letter [B] in front (for example, [B]309.2), are considered by the International Building Code Development Committee at the Code Development Hearing. Where this designation is applicable to the entire content of a main section of the code, the designation appears at the main section number and title and is not repeated at every subsection in that section.

The content of sections in this code which begin with a letter designation are maintained by another code development com- mittee in accordance with the following: [B]= International Building Code Development Committee; [E] = International Energy Conservation Code Development Committee; [EB] = International Existing Building Code Development Committee; [F] = International Fire Code Development Committee.

Marginal Markings

Solid vertical lines in the margins within the body of the code indicate a technical change from the requirements of the 2000 edition. Deletion indicators (➤) are provided in the margin where a paragraph or item has been deleted.

ORDINANCE

The *International Codes* are designed and promulgated to be adopted by reference by ordinance. Jurisdictions wishing to adopt the 2003 *International Plumbing Code* as an enforceable regulation governing plumbing systems should ensure that certain factual information is included in the adopting ordinance at the time adoption is being considered by the appropriate governmental body. The following sample adoption ordinance addresses several key elements of a code adoption ordinance, including the information required for insertion into the code text.

SAMPLE ORDINANCE FOR ADOPTION OF THE INTERNATIONAL PLUMBING CODE ORDINANCE NO._____

An ordinance of the [JURISDICTION] adopting the 2003 edition of the *International Plumbing Code*, regulating and governing the design, construction, quality of materials, erection, installation, alteration, repair, location, relocation, replacement, addition to, use or maintenance of plumbing systems in the [JURISDICTION]; providing for the issuance of permits and collection of fees therefor; repealing Ordinance No. _____ of the [JURISDICTION] and all other ordinances and parts of the ordinances in conflict therewith.

The [GOVERNING BODY] of the [JURISDICTION] does ordain as follows:

Section 1. That a certain document, three (3) copies of which are on file in the office of the [TITLE OF JURISDICTION'S KEEPER OF RECORDS] of [NAME OF JURISDICTION] , being marked and designated as the *International Plumbing Code*, 2003 edition, including Appendix Chapters [FILL IN THE APPENDIX CHAPTERS BEING ADOPTED], as published by the International Code Council, be and is hereby adopted as the Plumbing Code of the [JURISDICTION], in the State of [STATE NAME] regulating and governing the design, construction, quality of materials, erection, installation, alteration, repair, location, relocation, replacement, addition to, use or maintenance of plumbing systems as herein provided; providing for the issuance of permits and collection of fees therefor; and each and all of the regulations, provisions, penalties, conditions and terms of said Plumbing Code on file in the office of the [JURISDICTION] are hereby referred to, adopted, and made a part hereof, as if fully set out in this ordinance, with the additions, insertions, deletions and changes, if any, prescribed in Section 2 of this ordinance.

Section 2. The following sections are hereby revised:

Section 101.1. Insert: [NAME OF JURISDICTION]

Section 106.6.2. Insert: [APPROPRIATE SCHEDULE]

Section 106.6.3. Insert: [PERCENTAGES IN TWO LOCATIONS]

Section 108.4. Insert: [OFFENSE, DOLLAR AMOUNT, NUMBER OF DAYS]

Section 108.5. Insert: [DOLLAR AMOUNT IN TWO LOCATIONS]

Section 305.6.1. Insert: [NUMBER OF INCHES IN TWO LOCATIONS]

Section 904.1. Insert: [NUMBER OF INCHES]

Section 3. That Ordinance No. _____ of [JURISDICTION] entitled [FILL IN HERE THE COMPLETE TITLE OF THE ORDINANCE OR ORDINANCES IN EFFECT AT THE PRESENT TIME SO THAT THEY WILL BE REPEALED BY DEFINITE MENTION] and all other ordinances or parts of ordinances in conflict herewith are hereby repealed.

Section 4. That if any section, subsection, sentence, clause or phrase of this ordinance is, for any reason, held to be unconstitutional, such decision shall not affect the validity of the remaining portions of this ordinance. The [GOVERNING BODY] hereby declares that it would have passed this ordinance, and each section, subsection, clause or phrase thereof, irrespective of the fact that any one or more sections, subsections, sentences, clauses and phrases be declared unconstitutional.

Section 5. That nothing in this ordinance or in the Plumbing Code hereby adopted shall be construed to affect any suit or proceeding impending in any court, or any rights acquired, or liability incurred, or any cause or causes of action acquired or existing, under any act or ordinance hereby repealed as cited in Section 2 of this ordinance; nor shall any just or legal right or remedy of any character be lost, impaired or affected by this ordinance.

Section 6. That the [JURISDICTION'S KEEPER OF RECORDS] is hereby ordered and directed to cause this ordinance to be published. (An additional provision may be required to direct the number of times the ordinance is to be published and to specify that it is to be in a newspaper in general circulation. Posting may also be required.)

Section 7. That this ordinance and the rules, regulations, provisions, requirements, orders and matters established and adopted hereby shall take effect and be in full force and effect [TIME PERIOD] from and after the date of its final passage and adoption.

TABLE OF CONTENTS

CHAPTER 1

ADMINISTRATION

SECTION 101
GENERAL

101.1 Title. These regulations shall be known as the *International Plumbing Code* of [NAME OF JURISDICTION] hereinafter referred to as "this code."

101.2 Scope. The provisions of this code shall apply to the erection, installation, alteration, repairs, relocation, replacement, addition to, use or maintenance of plumbing systems within this jurisdiction. This code shall also regulate nonflammable medical gas, inhalation anesthetic, vacuum piping, nonmedical oxygen systems and sanitary and condensate vacuum collection systems. The installation of fuel gas distribution piping and equipment, fuel gas-fired water heaters and water heater venting systems shall be regulated by the *International Fuel Gas Code*. Provisions in the appendices shall not apply unless specifically adopted.

Exceptions:

1. Detached one- and two-family dwellings and multiple single-family dwellings (townhouses) not more than three stories high with separate means of egress and their accessory structures shall comply with the *International Residential Code.*

2. Plumbing systems in existing buildings undergoing repair, alteration, or additions, and change of occupancy shall be permitted to comply with the *International Existing Building Code.*

101.3 Intent. The purpose of this code is to provide minimum standards to safeguard life or limb, health, property and public welfare by regulating and controlling the design, construction, installation, quality of materials, location, operation and maintenance or use of plumbing equipment and systems.

101.4 Severability. If any section, subsection, sentence, clause or phrase of this code is for any reason held to be unconstitutional, such decision shall not affect the validity of the remaining portions of this code.

SECTION 102
APPLICABILITY

102.1 General. The provisions of this code shall apply to all matters affecting or relating to structures, as set forth in Section 101. Where, in any specific case, different sections of this code specify different materials, methods of construction or other requirements, the most restrictive shall govern.

102.2 Existing installations. Plumbing systems lawfully in existence at the time of the adoption of this code shall be permitted to have their use and maintenance continued if the use, maintenance or repair is in accordance with the original design and no hazard to life, health or property is created by such plumbing system.

102.3 Maintenance. All plumbing systems, materials and appurtenances, both existing and new, and all parts thereof, shall be maintained in proper operating condition in accordance with the original design in a safe and sanitary condition. All devices or safeguards required by this code shall be maintained in compliance with the code edition under which they were installed.

The owner or the owner's designated agent shall be responsible for maintenance of plumbing systems. To determine compliance with this provision, the code official shall have the authority to require any plumbing system to be reinspected.

[EB] 102.4 Additions, alterations or repairs. Additions, alterations, renovations or repairs to any plumbing system shall conform to that required for a new plumbing system without requiring the existing plumbing system to comply with all the requirements of this code. Additions, alterations or repairs shall not cause an existing system to become unsafe, insanitary or overloaded.

Minor additions, alterations, renovations and repairs to existing plumbing systems shall be permitted in the same manner and arrangement as in the existing system, provided that such repairs or replacement are not hazardous and are approved.

[EB] 102.5 Change in occupancy. It shall be unlawful to make any change in the occupancy of any structure that will subject the structure to any special provision of this code without approval of the code official. The code official shall certify that such structure meets the intent of the provisions of law governing building construction for the proposed new occupancy and that such change of occupancy does not result in any hazard to the public health, safety or welfare.

[EB] 102.6 Historic buildings. The provisions of this code relating to the construction, alteration, repair, enlargement, restoration, relocation or moving of buildings or structures shall not be mandatory for existing buildings or structures identified and classified by the state or local jurisdiction as historic buildings when such buildings or structures are judged by the code official to be safe and in the public interest of health, safety and welfare regarding any proposed construction, alteration, repair, enlargement, restoration, relocation or moving of buildings.

102.7 Moved buildings. Except as determined by Section 102.2, plumbing systems that are a part of buildings or structures moved into or within the jurisdiction shall comply with the provisions of this code for new installations.

102.8 Referenced codes and standards. The codes and standards referenced in this code shall be those that are listed in Chapter 13 and such codes and standards shall be considered as part of the requirements of this code to the prescribed extent of each such reference. Where the differences occur between provisions of this code and the referenced standards, the provisions of this code shall be the minimum requirements.

102.9 Requirements not covered by code. Any requirements necessary for the strength, stability or proper operation of an existing or proposed plumbing system, or for the public safety, health and general welfare, not specifically covered by this code shall be determined by the code official.

SECTION 103
DEPARTMENT OF PLUMBING INSPECTION

103.1 General. The department of plumbing inspection is hereby created and the executive official in charge thereof shall be known as the code official.

103.2 Appointment. The code official shall be appointed by the chief appointing authority of the jurisdiction, and the code official shall not be removed from office except for cause and after full opportunity to be heard on specific and relevant charges by and before the appointing authority.

103.3 Deputies. In accordance with the prescribed procedures of this jurisdiction and with the concurrence of the appointing authority, the code official shall have the authority to appoint a deputy code official, other related technical officers, inspectors and other employees.

103.4 Liability. The code official, officer or employee charged with the enforcement of this code, while acting for the jurisdiction, shall not thereby be rendered liable personally, and is hereby relieved from all personal liability for any damage accruing to persons or property as a result of any act required or permitted in the discharge of official duties.

Any suit instituted against any officer or employee because of an act performed by that officer or employee in the lawful discharge of duties and under the provisions of this code shall be defended by the legal representative of the jurisdiction until the final termination of the proceedings. The code official or any subordinate shall not be liable for costs in any action, suit or proceeding that is instituted in pursuance of the provisions of this code, and any officer of the department of plumbing inspection, acting in good faith and without malice, shall be free from liability for acts performed under any of its provisions or by reason of any act or omission in the performance of official duties in connection therewith.

SECTION 104
DUTIES AND POWERS OF THE CODE OFFICIAL

104.1 General. The code official shall enforce all of the provisions of this code and shall act on any question relative to the installation, alteration, repair, maintenance or operation of all plumbing systems, devices and equipment except as otherwise specifically provided for by statutory requirements or as provided for in Sections 104.2 through 104.8.

104.2 Rule-making authority. The code official shall have authority as necessary in the interest of public health, safety and general welfare to adopt and promulgate rules and regulations to interpret and implement the provisions of this code to secure the intent thereof and to designate requirements applicable because of local climatic or other conditions. Such rules shall not have the effect of waiving structural or fire performance requirements specifically provided for in this code, or of violating accepted engineering practice involving public safety.

104.3 Applications and permits. The code official shall receive applications and issue permits for the installation and alteration of plumbing, inspect the premises for which such permits have been issued, and enforce compliance with the provisions of this code.

104.4 Inspections. The code official shall make all the required inspections, or shall accept reports of inspection by approved agencies or individuals. All reports of such inspections shall be in writing and be certified by a responsible officer of such approved agency or by the responsible individual. The code official is authorized to engage such expert opinion as deemed necessary to report on unusual technical issues that arise, subject to the approval of the appointing authority.

104.5 Right of entry. Whenever it is necessary to make an inspection to enforce the provisions of this code, or whenever the code official has reasonable cause to believe that there exists in any building or upon any premises any conditions or violations of this code that make the building or premises unsafe, insanitary, dangerous or hazardous, the code official shall have the authority to enter the building or premises at all reasonable times to inspect or to perform the duties imposed upon the code official by this code. If such building or premises is occupied, the code official shall present credentials to the occupant and request entry. If such building or premises is unoccupied, the code official shall first make a reasonable effort to locate the owner or other person having charge or control of the building or premises and request entry. If entry is refused, the code official shall have recourse to every remedy provided by law to secure entry.

When the code official shall have first obtained a proper inspection warrant or other remedy provided by law to secure entry, no owner or occupant or person having charge, care or control of any building or premises shall fail or neglect, after proper request is made as herein provided, to promptly permit entry therein by the code official for the purpose of inspection and examination pursuant to this code.

104.6 Identification. The code official shall carry proper identification when inspecting structures or premises in the performance of duties under this code.

104.7 Notices and orders. The code official shall issue all necessary notices or orders to ensure compliance with this code.

104.8 Department records. The code official shall keep official records of applications received, permits and certificates issued, fees collected, reports of inspections, and notices and orders issued. Such records shall be retained in the official records as long as the building or structure to which such records relate remains in existence unless otherwise provided for by other regulations.

SECTION 105
APPROVAL

105.1 Modifications. Whenever there are practical difficulties involved in carrying out the provisions of this code, the code official shall have the authority to grant modifications for individual cases, provided the code official shall first find that special individual reason makes the strict letter of this code impractical and the modification is in conformity with the intent and purpose of this code and that such modification does not lessen health, life and fire safety requirements. The details of action granting modifications shall be recorded and entered in the files of the plumbing inspection department.

105.2 Alternative materials, methods and equipment. The provisions of this code are not intended to prevent the installation of any material or to prohibit any method of construction not specifically prescribed by this code, provided that any such alternative has been approved. An alternative material or method of construction shall be approved where the code official finds that the proposed design is satisfactory and complies with the intent of the provisions of this code, and that the material, method or work offered is, for the purpose intended, at least the equivalent of that prescribed in this code in quality, strength, effectiveness, fire resistance, durability and safety.

105.3 Required testing. Whenever there is insufficient evidence of compliance with the provisions of this code, or evidence that a material or method does not conform to the requirements of this code, or in order to substantiate claims for alternate materials or methods, the code official shall have the authority to require tests as evidence of compliance to be made at no expense to the jurisdiction.

105.3.1 Test methods. Test methods shall be as specified in this code or by other recognized test standards. In the absence of recognized and accepted test methods, the code official shall approve the testing procedures.

105.3.2 Testing agency. All tests shall be performed by an approved agency.

105.3.3 Test reports. Reports of tests shall be retained by the code official for the period required for retention of public records.

105.4 Alternative engineered design. The design, documentation, inspection, testing and approval of an alternative engineered design plumbing system shall comply with Sections 105.4.1 through 105.4.6.

105.4.1 Design criteria. An alternative engineered design shall conform to the intent of the provisions of this code and shall provide an equivalent level of quality, strength, effectiveness, fire resistance, durability and safety. Material, equipment or components shall be designed and installed in accordance with the manufacturer's installation instructions.

105.4.2 Submittal. The registered design professional shall indicate on the permit application that the plumbing system is an alternative engineered design. The permit and permanent permit records shall indicate that an alternative engineered design was part of the approved installation.

105.4.3 Technical data. The registered design professional shall submit sufficient technical data to substantiate the proposed alternative engineered design and to prove that the performance meets the intent of this code.

105.4.4 Construction documents. The registered design professional shall submit to the code official two complete sets of signed and sealed construction documents for the alternative engineered design. The construction documents shall include floor plans and a riser diagram of the work. Where appropriate, the construction documents shall indicate the direction of flow, all pipe sizes, grade of horizontal piping, loading, and location of fixtures and appliances.

105.4.5 Design approval. Where the code official determines that the alternative engineered design conforms to the intent of this code, the plumbing system shall be approved. If the alternative engineered design is not approved, the code official shall notify the registered design professional in writing, stating the reasons thereof.

105.4.6 Inspection and testing. The alternative engineered design shall be tested and inspected in accordance with the requirements of Sections 107 and 312.

105.5 Material and equipment reuse. Materials, equipment and devices shall not be reused unless such elements have been reconditioned, tested, placed in good and proper working condition and approved.

SECTION 106
PERMITS

106.1 When required. Any owner, authorized agent or contractor who desires to construct, enlarge, alter, repair, move, demolish or change the occupancy of a building or structure, or to erect, install, enlarge, alter, repair, remove, convert or replace any plumbing system, the installation of which is regulated by this code, or to cause any such work to be done, shall first make application to the code official and obtain the required permit for the work.

106.2 Exempt work. The following work shall be exempt from the requirement for a permit:

1. The stopping of leaks in drains, water, soil, waste or vent pipe provided, however, that if any concealed trap, drainpipe, water, soil, waste or vent pipe becomes defective and it becomes necessary to remove and replace the same with new material, such work shall be considered as new work and a permit shall be obtained and inspection made as provided in this code.

2. The clearing of stoppages or the repairing of leaks in pipes, valves or fixtures, and the removal and reinstallation of water closets, provided such repairs do not involve or require the replacement or rearrangement of valves, pipes or fixtures.

Exemption from the permit requirements of this code shall not be deemed to grant authorization for any work to be done in violation of the provisions of this code or any other laws or ordinances of this jurisdiction.

106.3 Application for permit. Each application for a permit, with the required fee, shall be filed with the code official on a form furnished for that purpose and shall contain a general description of the proposed work and its location. The application shall be signed by the owner or an authorized agent. The permit application shall indicate the proposed occupancy of all parts of the building and of that portion of the site or lot, if any, not covered by the building or structure and shall contain such other information required by the code official.

106.3.1 Construction documents. Construction documents, engineering calculations, diagrams and other such data shall be submitted in two or more sets with each application for a permit. The code official shall require construction documents, computations and specifications to be prepared and designed by a registered design professional when required by state law. Construction documents shall be drawn to scale and shall be of sufficient clarity to indicate the location, nature and extent of the work proposed and show in detail that the work conforms to the provisions of this code. Construction documents for buildings more than two stories in height shall indicate where penetrations will be made for pipe, fittings and components and shall indicate the materials and methods for maintaining required structural safety, fire-resistance rating and fireblocking.

> **Exception:** The code official shall have the authority to waive the submission of construction documents, calculations or other data if the nature of the work applied for is such that reviewing of construction documents is not necessary to determine compliance with this code.

106.4 By whom application is made. Application for a permit shall be made by the person or agent to install all or part of any plumbing system. The applicant shall meet all qualifications established by statute, or by rules promulgated by this code, or by ordinance or by resolution. The full name and address of the applicant shall be stated in the application.

106.5 Permit issuance. The application, construction documents and other data filed by an applicant for permit shall be reviewed by the code official. If the code official finds that the proposed work conforms to the requirements of this code and all laws and ordinances applicable thereto, and that the fees specified in Section 106.6 have been paid, a permit shall be issued to the applicant.

106.5.1 Approved construction documents. When the code official issues the permit where construction documents are required, the construction documents shall be endorsed in writing and stamped "APPROVED." Such approved construction documents shall not be changed, modified or altered without authorization from the code official. All work shall be done in accordance with the approved construction documents.

The code official shall have the authority to issue a permit for the construction of a part of a plumbing system before the entire construction documents for the whole system have been submitted or approved, provided adequate information and detailed statements have been filed complying with all pertinent requirements of this code. The holders of such permit shall proceed at their own risk without assurance that the permit for the entire plumbing system will be granted.

106.5.2 Validity. The issuance of a permit or approval of construction documents shall not be construed to be a permit for, or an approval of, any violation of any of the provisions of this code or any other ordinance of the jurisdiction. No permit presuming to give authority to violate or cancel the provisions of this code shall be valid.

The issuance of a permit based upon construction documents and other data shall not prevent the code official from thereafter requiring the correction of errors in said construction documents and other data or from preventing building operations being carried on thereunder when in violation of this code or of other ordinances of this jurisdiction.

106.5.3 Expiration. Every permit issued by the code official under the provisions of this code shall expire by limitation and become null and void if the work authorized by such permit is not commenced within 180 days from the date of such permit, or if the work authorized by such permit is suspended or abandoned at any time after the work is commenced for a period of 180 days. Before such work can be recommenced, a new permit shall be first obtained and the fee therefor shall be one-half the amount required for a new permit for such work, provided no changes have been made or will be made in the original construction documents for such work, and provided further that such suspension or abandonment has not exceeded 1 year.

106.5.4 Extensions. Any permittee holding an unexpired permit shall have the right to apply for an extension of the time within which the permittee will commence work under that permit when work is unable to be commenced within the time required by this section for good and satisfactory reasons. The code official shall extend the time for action by the permittee for a period not exceeding 180 days if there is reasonable cause. No permit shall be extended more than once. The fee for an extension shall be one-half the amount required for a new permit for such work.

106.5.5 Suspension or revocation of permit. The code official shall revoke a permit or approval issued under the provisions of this code in case of any false statement or misrepresentation of fact in the application or on the construction documents upon which the permit or approval was based.

106.5.6 Retention of construction documents. One set of construction documents shall be retained by the code official until final approval of the work covered therein. One set of approved construction documents shall be returned to the applicant, and said set shall be kept on the site of the building or work at all times during which the work authorized thereby is in progress.

106.6 Fees. A permit shall not be issued until the fees prescribed in Section 106.6.2 have been paid, and an amendment to a permit shall not be released until the additional fee, if any, due to an increase of the plumbing systems, has been paid.

106.6.1 Work commencing before permit issuance. Any person who commences any work on a plumbing system before obtaining the necessary permits shall be subject to 100 percent of the usual permit fee in addition to the required permit fees.

106.6.2 Fee schedule. The fees for all plumbing work shall be as indicated in the following schedule:

[JURISDICTION TO INSERT APPROPRIATE SCHEDULE]

106.6.3 Fee refunds. The code official shall authorize the refunding of fees as follows:

1. The full amount of any fee paid hereunder that was erroneously paid or collected.

2. Not more than [SPECIFY PERCENTAGE] percent of the permit fee paid when no work has been done under a permit issued in accordance with this code.

3. Not more than [SPECIFY PERCENTAGE] percent of the plan review fee paid when an application for a permit for which a plan review fee has been paid is withdrawn

or canceled before any plan review effort has been expended.

The code official shall not authorize the refunding of any fee paid except upon written application filed by the original permittee not later than 180 days after the date of fee payment.

SECTION 107
INSPECTIONS AND TESTING

107.1 Required inspections and testing. The code official, upon notification from the permit holder or the permit holder's agent, shall make the following inspections and such other inspections as necessary, and shall either release that portion of the construction or shall notify the permit holder or an agent of any violations that must be corrected. The holder of the permit shall be responsible for the scheduling of such inspections.

1. Underground inspection shall be made after trenches or ditches are excavated and bedded, piping installed, and before any backfill is put in place.

2. Rough-in inspection shall be made after the roof, framing, fireblocking, firestopping, draftstopping and bracing is in place and all sanitary, storm and water distribution piping is roughed-in, and prior to the installation of wall or ceiling membranes.

3. Final inspection shall be made after the building is complete, all plumbing fixtures are in place and properly connected, and the structure is ready for occuancy.

107.1.1 Approved agencies. Test reports submitted to the code official for consideration shall be developed by approved agencies that have satisfied the requirements as to qualifications and reliability.

107.1.2 Evaluation and follow-up inspection services. Prior to the approval of a closed, prefabricated plumbing system and the issuance of a plumbing permit, the code official shall require the submittal of an evaluation report on each prefabricated plumbing system indicating the complete details of the plumbing system, including a description of the system and its components, the basis upon which the plumbing system is being evaluated, test results and similar information, and other data as necessary for the code official to determine conformance to this code.

107.1.2.1 Evaluation service. The code official shall designate the evaluation service of an approved agency as the evaluation agency, and review such agency's evaluation report for adequacy and conformance to this code.

107.1.2.2 Follow-up inspection. Except where ready access is provided to all plumbing systems, service equipment and accessories for complete inspection at the site without disassembly or dismantling, the code official shall conduct the frequency of in-plant inspections necessary to ensure conformance to the approved evaluation report or shall designate an independent, approved inspection agency to conduct such inspections. The inspection agency shall furnish the code official with the follow-up inspection manual and a report of inspections upon request, and the plumbing system shall have an identifying label permanently affixed to the system indicating that factory inspections have been performed.

107.1.2.3 Test and inspection records. All required test and inspection records shall be available to the code official at all times during the fabrication of the plumbing system and the erection of the building, or such records as the code official designates shall be filed.

107.2 Special inspections. Special inspections of alternative engineered design plumbing systems shall be conducted in accordance with Sections 107.2.1 and 107.2.2.

107.2.1 Periodic inspection. The registered design professional or designated inspector shall periodically inspect and observe the alternative engineered design to determine that the installation is in accordance with the approved construction documents. All discrepancies shall be brought to the immediate attention of the plumbing contractor for correction. Records shall be kept of all inspections.

107.2.2 Written report. The registered design professional shall submit a final report in writing to the code official upon completion of the installation, certifying that the alternative engineered design conforms to the approved construction documents. A notice of approval for the plumbing system shall not be issued until a written certification has been submitted.

107.3 Testing. Plumbing work and systems shall be tested as required in Section 312 and in accordance with Sections 107.3.1 through 107.3.3. Tests shall be made by the permit holder and observed by the code official.

107.3.1 New, altered, extended or repaired systems. New plumbing systems and parts of existing systems that have been altered, extended or repaired shall be tested as prescribed herein to disclose leaks and defects, except that testing is not required in the following cases:

1. In any case that does not include addition to, replacement, alteration or relocation of any water supply, drainage or vent piping.

2. In any case where plumbing equipment is set up temporarily for exhibition purposes.

107.3.2 Equipment, material and labor for tests. All equipment, material and labor required for testing a plumbing system or part thereof shall be furnished by the permit holder.

107.3.3 Reinspection and testing. Where any work or installation does not pass any initial test or inspection, the necessary corrections shall be made to comply with this code. The work or installation shall then be resubmitted to the code official for inspection and testing.

107.4 Approval. After the prescribed tests and inspections indicate that the work complies in all respects with this code, a notice of approval shall be issued by the code official.

107.5 Temporary connection. The code official shall have the authority to authorize the temporary connection of the building or system to the utility source for the purpose of testing plumbing systems or for use under a temporary certificate of occupancy.

SECTION 108
VIOLATIONS

108.1 Unlawful acts. It shall be unlawful for any person, firm or corporation to erect, construct, alter, repair, remove, demolish or utilize any plumbing system, or cause same to be done, in conflict with or in violation of any of the provisions of this code.

108.2 Notice of violation. The code official shall serve a notice of violation or order to the person responsible for the erection, installation, alteration, extension, repair, removal or demolition of plumbing work in violation of the provisions of this code, or in violation of a detail statement or the approved construction documents thereunder, or in violation of a permit or certificate issued under the provisions of this code. Such order shall direct the discontinuance of the illegal action or condition and the abatement of the violation.

108.3 Prosecution of violation. If the notice of violation is not complied with promptly, the code official shall request the legal counsel of the jurisdiction to institute the appropriate proceeding at law or in equity to restrain, correct or abate such violation, or to require the removal or termination of the unlawful occupancy of the structure in violation of the provisions of this code or of the order or direction made pursuant thereto.

108.4 Violation penalties. Any person who shall violate a provision of this code or shall fail to comply with any of the requirements thereof or who shall erect, install, alter or repair plumbing work in violation of the approved construction documents or directive of the code official, or of a permit or certificate issued under the provisions of this code, shall be guilty of a [SPECIFY OFFENSE], punishable by a fine of not more than [AMOUNT] dollars or by imprisonment not exceeding [NUMBER OF DAYS], or both such fine and imprisonment. Each day that a violation continues after due notice has been served shall be deemed a separate offense.

108.5 Stop work orders. Upon notice from the code official, work on any plumbing system that is being done contrary to the provisions of this code or in a dangerous or unsafe manner shall immediately cease. Such notice shall be in writing and shall be given to the owner of the property, or to the owner's agent, or to the person doing the work. The notice shall state the conditions under which work is authorized to resume. Where an emergency exists, the code official shall not be required to give a written notice prior to stopping the work. Any person who shall continue any work in or about the structure after having been served with a stop work order, except such work as that person is directed to perform to remove a violation or unsafe condition, shall be liable to a fine of not less than [AMOUNT] dollars or more than [AMOUNT] dollars.

108.6 Abatement of violation. The imposition of the penalties herein prescribed shall not preclude the legal officer of the jurisdiction from instituting appropriate action to prevent unlawful construction or to restrain, correct or abate a violation, or to prevent illegal occupancy of a building, structure or premises, or to stop an illegal act, conduct, business or utilization of the plumbing on or about any premises.

108.7 Unsafe plumbing. Any plumbing regulated by this code that is unsafe or that constitutes a fire or health hazard, insanitary condition, or is otherwise dangerous to human life is hereby declared unsafe. Any use of plumbing regulated by this code constituting a hazard to safety, health or public welfare by reason of inadequate maintenance, dilapidation, obsolescence, fire hazard, disaster, damage or abandonment is hereby declared an unsafe use. Any such unsafe equipment is hereby declared to be a public nuisance and shall be abated by repair, rehabilitation, demolition or removal.

108.7.1 Authority to condemn equipment. Whenever the code official determines that any plumbing, or portion thereof, regulated by this code has become hazardous to life, health or property or has become insanitary, the code official shall order in writing that such plumbing either be removed or restored to a safe or sanitary condition. A time limit for compliance with such order shall be specified in the written notice. No person shall use or maintain defective plumbing after receiving such notice.

When such plumbing is to be disconnected, written notice as prescribed in Section 108.2 shall be given. In cases of immediate danger to life or property, such disconnection shall be made immediately without such notice.

108.7.2 Authority to disconnect service utilities. The code official shall have the authority to authorize disconnection of utility service to the building, structure or system regulated by the technical codes in case of an emergency, where necessary, to eliminate an immediate danger to life or property. Where possible, the owner and occupant of the building, structure or service system shall be notified of the decision to disconnect utility service prior to taking such action. If not notified prior to disconnecting, the owner or occupant of the building, structure or service systems shall be notified in writing, as soon as practical thereafter.

108.7.3 Connection after order to disconnect. No person shall make connections from any energy, fuel, power supply or water distribution system or supply energy, fuel or water to any equipment regulated by this code that has been disconnected or ordered to be disconnected by the code official or the use of which has been ordered to be discontinued by the code official until the code official authorizes the reconnection and use of such equipment.

When any plumbing is maintained in violation of this code, and in violation of any notice issued pursuant to the provisions of this section, the code official shall institute any appropriate action to prevent, restrain, correct or abate the violation.

SECTION 109
MEANS OF APPEAL

109.1 Application for appeal. Any person shall have the right to appeal a decision of the code official to the board of appeals. An application for appeal shall be based on a claim that the true intent of this code or the rules legally adopted thereunder have been incorrectly interpreted, the provisions of this code do not fully apply, or an equally good or better form of construction is proposed. The application shall be filed on a form obtained from the code official within 20 days after the notice was served.

109.2 Membership of board. The board of appeals shall consist of five members appointed by the chief appointing authority as follows: one for 5 years, one for 4 years, one for 3 years, one for 2 years and one for 1 year. Thereafter, each new member shall serve for 5 years or until a successor has been appointed.

109.2.1 Qualifications. The board of appeals shall consist of five individuals, one from each of the following professions or disciplines:

1. Registered design professional who is a registered architect; or a builder or superintendent of building construction with at least 10 years' experience, 5 years of which shall have been in responsible charge of work.

2. Registered design professional with structural engineering or architectural experience.

3. Registered design professional with mechanical and plumbing engineering experience; or a mechanical and plumbing contractor with at least 10 years' experience, 5 years of which shall have been in responsible charge of work.

4. Registered design professional with electrical engineering experience; or an electrical contractor with at least 10 years' experience, 5 years of which shall have been in responsible charge of work.

5. Registered design professional with fire protection engineering experience; or a fire protection contractor with at least 10 years' experience, 5 years of which shall have been in responsible charge of work.

109.2.2 Alternate members. The chief appointing authority shall appoint two alternate members who shall be called by the board chairman to hear appeals during the absence or disqualification of a member. Alternate members shall possess the qualifications required for board membership, and shall be appointed for 5 years or until a successor has been appointed.

109.2.3 Chairman. The board shall annually select one of its members to serve as chairman.

109.2.4 Disqualification of member. A member shall not hear an appeal in which that member has any personal, professional or financial interest.

109.2.5 Secretary. The chief administrative officer shall designate a qualified clerk to serve as secretary to the board. The secretary shall file a detailed record of all proceedings in the office of the chief administrative officer.

109.2.6 Compensation of members. Compensation of members shall be determined by law.

109.3 Notice of meeting. The board shall meet upon notice from the chairman, within 10 days of the filing of an appeal or at stated periodic meetings.

109.4 Open hearing. All hearings before the board shall be open to the public. The appellant, the appellant's representative, the code official and any person whose interests are affected shall be given an opportunity to be heard.

109.4.1 Procedure. The board shall adopt and make available to the public through the secretary procedures under which a hearing will be conducted. The procedures shall not require compliance with strict rules of evidence, but shall mandate that only relevant information be received.

109.5 Postponed hearing. When five members are not present to hear an appeal, either the appellant or the appellant's representative shall have the right to request a postponement of the hearing.

109.6 Board decision. The board shall modify or reverse the decision of the code official by a concurring vote of three members.

109.6.1 Resolution. The decision of the board shall be by resolution. Certified copies shall be furnished to the appellant and to the code official.

109.6.2 Administration. The code official shall take immediate action in accordance with the decision of the board.

109.7 Court review. Any person, whether or not a previous party of the appeal, shall have the right to apply to the appropriate court for a writ of certiorari to correct errors of law. Application for review shall be made in the manner and time required by law following the filing of the decision in the office of the chief administrative officer.

CHAPTER 2

DEFINITIONS

SECTION 201
GENERAL

201.1 Scope. Unless otherwise expressly stated, the following words and terms shall, for the purposes of this code, have the meanings shown in this chapter.

201.2 Interchangeability. Words stated in the present tense include the future; words stated in the masculine gender include the feminine and neuter; the singular number includes the plural and the plural the singular.

201.3 Terms defined in other codes. Where terms are not defined in this code and are defined in the *International Building Code, International Fire Code,* ICC *Electrical Code, International Fuel Gas Code* or the *International Mechanical Code,* such terms shall have the meanings ascribed to them as in those codes.

201.4 Terms not defined. Where terms are not defined through the methods authorized by this section, such terms shall have ordinarily accepted meanings such as the context implies.

SECTION 202
GENERAL DEFINITIONS

ACCEPTED ENGINEERING PRACTICE. That which conforms to accepted principles, tests or standards of nationally recognized technical or scientific authorities.

ACCESS (TO). That which enables a fixture, appliance or equipment to be reached by ready access or by a means that first requires the removal or movement of a panel, door or similar obstruction (see "Ready access").

ACCESS COVER. A removable plate, usually secured by bolts or screws, to permit access to a pipe or pipe fitting for the purposes of inspection, repair or cleaning.

ADAPTER FITTING. An approved connecting device that suitably and properly joins or adjusts pipes and fittings which do not otherwise fit together.

AIR ADMITTANCE VALVE. One-way valve designed to allow air to enter the plumbing drainage system when negative pressures develop in the piping system. The device shall close by gravity and seal the vent terminal at zero differential pressure (no flow conditions) and under positive internal pressures. The purpose of an air admittance valve is to provide a method of allowing air to enter the plumbing drainage system without the use of a vent extended to open air and to prevent sewer gases from escaping into a building.

AIR BREAK (Drainage System). A piping arrangement in which a drain from a fixture, appliance or device discharges indirectly into another fixture, receptacle or interceptor at a point below the flood level rim and above the trap seal.

AIR GAP (Drainage System). The unobstructed vertical distance through the free atmosphere between the outlet of the waste pipe and the flood level rim of the receptacle into which the waste pipe is discharging.

AIR GAP (Water Distribution System). The unobstructed vertical distance through the free atmosphere between the lowest opening from any pipe or faucet supplying water to a tank, plumbing fixture or other device and the flood level rim of the receptacle.

ALTERNATIVE ENGINEERED DESIGN. A plumbing system that performs in accordance with the intent of Chapters 3 through 12 and provides an equivalent level of performance for the protection of public health, safety and welfare. The system design is not specifically regulated by Chapters 3 through 12.

ANCHORS. See "Supports."

ANTISIPHON. A term applied to valves or mechanical devices that eliminate siphonage.

APPROVED. Approved by the code official or other authority having jurisdiction.

APPROVED AGENCY. An established and recognized agency approved by the code official and that is regularly engaged in conducting tests or furnishing inspection services.

AREA DRAIN. A receptacle designed to collect surface or storm water from an open area.

ASPIRATOR. A fitting or device supplied with water or other fluid under positive pressure that passes through an integral orifice or constriction, causing a vacuum. Aspirators are also referred to as suction apparatus, and are similar in operation to an ejector.

BACKFLOW. Pressure created by any means in the water distribution system, which by being in excess of the pressure in the water supply mains causes a potential backflow condition.

> **Backpressure, low head.** A pressure less than or equal to 4.33 psi (29.88 kPa) or the pressure exerted by a 10-foot (3048 mm) column of water.

> **Backsiphonage.** The backflow of potentially contaminated water into the potable water system as a result of the pressure in the potable water system falling below atmospheric pressure of the plumbing fixtures, pools, tanks or vats connected to the potable water distribution piping.

> **Backwater valve.** A device or valve installed in the building drain or sewer pipe where a sewer is subject to backflow, and which prevents drainage or waste from backing up into a low level or fixtures and causing a flooding condition.

> **Drainage.** A reversal of flow in the drainage system.

> **Water supply system.** The flow of water or other liquids, mixtures or substances into the distribution pipes of a potable water supply from any source except the intended source.

BACKFLOW CONNECTION. Any arrangement whereby backflow is possible.

BACKFLOW PREVENTER. A device or means to prevent backflow.

BALL COCK. See "Fill Valve."

BASE FLOOD ELEVATION. A reference point, determined in accordance with the building code, based on the depth or peak elevation of flooding, including wave height, which has a 1 percent (100-year flood) or greater chance of occurring in any given year.

BATHROOM GROUP. A group of fixtures consisting of a water closet, lavatory, bathtub or shower, including or excluding a bidet, an emergency floor drain or both. Such fixtures are located together on the same floor level.

BEDPAN STEAMER OR BOILER. A fixture utilized for scalding bedpans or urinals by direct application of steam or boiling water.

BEDPAN WASHER AND STERILIZER. A fixture designed to wash bedpans and to flush the contents into the sanitary drainage system. Included are fixtures of this type that provide for disinfecting utensils by scalding with steam or hot water.

BEDPAN WASHER HOSE. A device supplied with hot and cold water and located adjacent to a water closet or clinical sink to be utilized for cleansing bedpans.

BRANCH. Any part of the piping system except a riser, main or stack.

BRANCH INTERVAL. A distance along a soil or waste stack corresponding in general to a story height, but not less than 8 feet (2438 mm), within which the horizontal branches from one floor or story of a structure are connected to the stack.

BRANCH VENT. A vent connecting one or more individual vents with a vent stack or stack vent.

BUILDING. Any structure occupied or intended for supporting or sheltering any occupancy.

BUILDING DRAIN. That part of the lowest piping of a drainage system that receives the discharge from soil, waste and other drainage pipes inside and that extends 30 inches (762 mm) in developed length of pipe beyond the exterior walls of the building and conveys the drainage to the building sewer.

> **Combined.** A building drain that conveys both sewage and storm water or other drainage.

> **Sanitary.** A building drain that conveys sewage only.

> **Storm.** A building drain that conveys storm water or other drainage, but not sewage.

BUILDING SEWER. That part of the drainage system that extends from the end of the building drain and conveys the discharge to a public sewer, private sewer, individual sewage disposal system or other point of disposal.

> **Combined.** A building sewer that conveys both sewage and storm water or other drainage.

> **Sanitary.** A building sewer that conveys sewage only.

> **Storm.** A building sewer that conveys storm water or other drainage, but not sewage.

BUILDING SUBDRAIN. That portion of a drainage system that does not drain by gravity into the building sewer.

BUILDING TRAP. A device, fitting or assembly of fittings installed in the building drain to prevent circulation of air between the drainage system of the building and the building sewer.

CIRCUIT VENT. A vent that connects to a horizontal drainage branch and vents two traps to a maximum of eight traps or trapped fixtures connected into a battery.

CISTERN. A small covered tank for storing water for a home or farm. Generally, this tank stores rainwater to be utilized for purposes other than in the potable water supply, and such tank is placed underground in most cases.

CLEANOUT. An access opening in the drainage system utilized for the removal of obstructions. Types of cleanouts include a removable plug or cap, and a removable fixture or fixture trap.

CODE. These regulations, subsequent amendments thereto, or any emergency rule or regulation that the administrative authority having jurisdiction has lawfully adopted.

CODE OFFICIAL. The officer or other designated authority charged with the administration and enforcement of this code, or a duly authorized representative.

COMBINATION FIXTURE. A fixture combining one sink and laundry tray or a two- or three-compartment sink or laundry tray in one unit.

COMBINATION WASTE AND VENT SYSTEM. A specially designed system of waste piping embodying the horizontal wet venting of one or more sinks or floor drains by means of a common waste and vent pipe adequately sized to provide free movement of air above the flow line of the drain.

COMBINED BUILDING DRAIN. See "Building drain, combined."

COMBINED BUILDING SEWER. See "Building sewer, combined."

COMMON VENT. A vent connecting at the junction of two fixture drains or to a fixture branch and serving as a vent for both fixtures.

CONCEALED FOULING SURFACE. Any surface of a plumbing fixture which is not readily visible and is not scoured or cleansed with each fixture operation.

CONDUCTOR. A pipe inside the building that conveys storm water from the roof to a storm or combined building drain.

CONSTRUCTION DOCUMENTS. All of the written, graphic and pictorial documents prepared or assembled for describing the design, location and physical characteristics of the elements of the project necessary for obtaining a building permit. The construction drawings shall be drawn to an appropriate scale.

CONTAMINATION. An impairment of the quality of the potable water that creates an actual hazard to the public health through poisoning or through the spread of disease by sewage, industrial fluids or waste.

CRITICAL LEVEL (C-L). An elevation (height) reference point that determines the minimum height at which a backflow preventer or vacuum breaker is installed above the flood level rim of the fixture or receptor served by the device. The critical level is the elevation level below which there is a potential for backflow to occur. If the critical level marking is not indicated on the device, the bottom of the device shall constitute the critical level.

CROSS CONNECTION. Any physical connection or arrangement between two otherwise separate piping systems, one of which contains potable water and the other either water of unknown or questionable safety or steam, gas or chemical, whereby there exists the possibility for flow from one system to the other, with the direction of flow depending on the pressure differential between the two systems (see "Backflow").

DEAD END. A branch leading from a soil, waste or vent pipe; a building drain; or a building sewer, and terminating at a developed length of 2 feet (610 mm) or more by means of a plug, cap or other closed fitting.

DEPTH OF WATER SEAL. The depth of water that would have to be removed from a full trap before air could pass through the trap.

DESIGN FLOOD ELEVATION. The elevation of the "design flood," including wave height, relative to the datum specified on the community's legally designated flood hazard map.

DEVELOPED LENGTH. The length of a pipeline measured along the centerline of the pipe and fittings.

DISCHARGE PIPE. A pipe that conveys the discharges from plumbing fixtures or appliances.

DRAIN. Any pipe that carries wastewater or water-borne wastes in a building drainage system.

DRAINAGE FITTINGS. Type of fitting or fittings utilized in the drainage system. Drainage fittings are similar to cast-iron fittings, except that instead of having a bell and spigot, drainage fittings are recessed and tapped to eliminate ridges on the inside of the installed pipe.

DRAINAGE FIXTURE UNIT

Drainage (dfu). A measure of the probable discharge into the drainage system by various types of plumbing fixtures. The drainage fixture-unit value for a particular fixture depends on its volume rate of drainage discharge, on the time duration of a single drainage operation and on the average time between successive operations.

DRAINAGE SYSTEM. Piping within a public or private premise that conveys sewage, rainwater or other liquid wastes to a point of disposal. A drainage system does not include the mains of a public sewer system or a private or public sewage treatment or disposal plant.

Building gravity. A drainage system that drains by gravity into the building sewer.

Sanitary. A drainage system that carries sewage and excludes storm, surface and ground water.

Storm. A drainage system that carries rainwater, surface water, subsurface water and similar liquid wastes.

EFFECTIVE OPENING. The minimum cross-sectional area at the point of water supply discharge, measured or expressed in terms of the diameter of a circle or, if the opening is not circular, the diameter of a circle of equivalent cross-sectional area. For faucets and similar fittings, the effective opening shall be measured at the smallest orifice in the fitting body or in the supply piping to the fitting.

EMERGENCY FLOOR DRAIN. A floor drain that does not receive the discharge of any drain or indirect waste pipe, and that protects against damage from accidental spills, fixture overflows and leakage.

ESSENTIALLY NONTOXIC TRANSFER FLUIDS. Fluids having a Gosselin rating of 1, including propylene glycol; mineral oil; polydimethylsiloxane; hydrochlorofluorocarbon, chlorofluorocarbon and carbon refrigerants; and FDA-approved boiler water additives for steam boilers.

ESSENTIALLY TOXIC TRANSFER FLUIDS. Soil, waste or gray water and fluids having a Gosselin rating of 2 or more including ethylene glycol, hydrocarbon oils, ammonia refrigerants and hydrazine.

EXISTING INSTALLATIONS. Any plumbing system regulated by this code that was legally installed prior to the effective date of this code, or for which a permit to install has been issued.

FAUCET. A valve end of a water pipe through which water is drawn from or held within the pipe.

FILL VALVE. A water supply valve, opened or closed by means of a float or similar device, utilized to supply water to a tank. An antisiphon fill valve contains an antisiphon device in the form of an approved air gap or vacuum breaker that is an integral part of the fill valve unit and that is positioned on the discharge side of the water supply control valve.

FIXTURE. See "Plumbing fixture."

FIXTURE BRANCH. A drain serving two or more fixtures that discharges to another drain or to a stack.

FIXTURE DRAIN. The drain from the trap of a fixture to a junction with any other drain pipe.

FIXTURE FITTING

Supply fitting. A fitting that controls the volume and/or directional flow of water and is either attached to or accessible from a fixture, or is used with an open or atmospheric discharge.

Waste fitting. A combination of components that conveys the sanitary waste from the outlet of a fixture to the connection to the sanitary drainage system.

FIXTURE SUPPLY. The water supply pipe connecting a fixture to a branch water supply pipe or directly to a main water supply pipe.

FLOOD LEVEL RIM. The edge of the receptacle from which water overflows.

FLOOD HAZARD AREA. The greater of the following two areas:

1. The area within a flood plain subject to a 1-percent or greater chance of flooding in any given year.

2. The area designated as a flood hazard area on a community's flood hazard map or as otherwise legally designated.

FLOW PRESSURE. The pressure in the water supply pipe near the faucet or water outlet while the faucet or water outlet is wide open and flowing.

FLUSH TANK. A tank designed with a ball cock and flush valve to flush the contents of the bowl or usable portion of the fixture.

FLUSHOMETER TANK. A device integrated within an air accumulator vessel that is designed to discharge a predetermined quantity of water to fixtures for flushing purposes.

FLUSHOMETER VALVE. A valve attached to a pressurized water supply pipe and so designed that when activated it opens the line for direct flow into the fixture at a rate and quantity to operate the fixture properly, and then gradually closes to reseal fixture traps and avoid water hammer.

GREASE INTERCEPTOR. A passive interceptor whose rated flow exceeds 50 gpm (189 L/m).

GREASE-LADEN WASTE. Effluent discharge that is produced from food processing, food preparation or other sources where grease, fats and oils enter automatic dishwater prerinse stations, sinks or other appurtenances.

GREASE TRAP. A passive interceptor whose rated flow is 50 gpm (189 L/m) or less.

HANGERS. See "Supports."

HORIZONTAL BRANCH DRAIN. A drainage branch pipe extending laterally from a soil or waste stack or building drain, with or without vertical sections or branches, that receives the discharge from two or more fixture drains or branches and conducts the discharge to the soil or waste stack or to the building drain.

HORIZONTAL PIPE. Any pipe or fitting that makes an angle of less than 45 degrees (0.79 rad) with the horizontal.

HOT WATER. Water at a temperature greater than or equal to 110°F (43°C).

HOUSE TRAP. See "Building trap."

INDIRECT WASTE PIPE. A waste pipe that does not connect directly with the drainage system, but that discharges into the drainage system through an air break or air gap into a trap, fixture, receptor or interceptor.

INDIVIDUAL SEWAGE DISPOSAL SYSTEM. A system for disposal of domestic sewage by means of a septic tank, cesspool or mechanical treatment, designed for utilization apart from a public sewer to serve a single establishment or building.

INDIVIDUAL VENT. A pipe installed to vent a fixture trap and connects with the vent system above the fixture served or terminates in the open air.

INDIVIDUAL WATER SUPPLY. A water supply that serves one or more families, and that is not an approved public water supply.

INTERCEPTOR. A device designed and installed to separate and retain for removal, by automatic or manual means, deleterious, hazardous or undesirable matter from normal wastes, while permitting normal sewage or wastes to discharge into the drainage system by gravity.

JOINT

    Expansion. A loop, return bend or return offset that provides for the expansion and contraction in a piping system and is utilized in tall buildings or where there is a rapid change of temperature, as in power plants, steam rooms and similar occupancies.

    Flexible. Any joint between two pipes that permits one pipe to be deflected or moved without movement or deflection of the other pipe.

    Mechanical. See "Mechanical joint."

    Slip. A type of joint made by means of a washer or a special type of packing compound in which one pipe is slipped into the end of an adjacent pipe.

LEAD-FREE PIPE AND FITTINGS. Containing not more than 8.0-percent lead.

LEAD-FREE SOLDER AND FLUX. Containing not more than 0.2-percent lead.

LEADER. An exterior drainage pipe for conveying storm water from roof or gutter drains to an approved means of disposal.

LOCAL VENT STACK. A vertical pipe to which connections are made from the fixture side of traps and through which vapor or foul air is removed from the fixture or device utilized on bedpan washers.

MACERATING TOILET SYSTEMS. An assembly consisting of a water closet and sump with a macerating pump that is designed to collect, grind and pump wastes from the water closet and up to two other fixtures connected to the sump.

MAIN. The principal pipe artery to which branches are connected.

MANIFOLD. See "Plumbing appurtenance."

MECHANICAL JOINT. A connection between pipes, fittings, or pipes and fittings that is not screwed, caulked, threaded, soldered, solvent cemented, brazed or welded. A joint in which compression is applied along the centerline of the pieces being joined. In some applications, the joint is part of a coupling, fitting or adapter.

MEDICAL GAS SYSTEM. The complete system to convey medical gases for direct patient application from central supply systems (bulk tanks, manifolds and medical air compressors), with pressure and operating controls, alarm warning systems, related components and piping networks extending to station outlet valves at patient use points.

MEDICAL VACUUM SYSTEMS. A system consisting of central-vacuum-producing equipment with pressure and operating controls, shutoff valves, alarm-warning systems, gauges and a network of piping extending to and terminating with suitable station inlets at locations where patient suction may be required.

NONPOTABLE WATER. Water not safe for drinking, personal or culinary utilization.

NUISANCE. Public nuisance as known in common law or in equity jurisprudence; whatever is dangerous to human life or detrimental to health; whatever structure or premises is not sufficiently ventilated, sewered, drained, cleaned or lighted, with respect to its intended occupancy; and whatever renders the air, or human food, drink or water supply unwholesome.

OCCUPANCY. The purpose for which a building or portion thereof is utilized or occupied.

OFFSET. A combination of approved bends that makes two changes in direction bringing one section of the pipe out of line but into a line parallel with the other section.

OPEN AIR. Outside the structure.

PLUMBING. The practice, materials and fixtures utilized in the installation, maintenance, extension and alteration of all piping, fixtures, plumbing appliances and plumbing appurtenances, within or adjacent to any structure, in connection with sanitary drainage or storm drainage facilities; venting systems; and public or private water supply systems.

PLUMBING APPLIANCE. Any one of a special class of plumbing fixtures intended to perform a special function. Included are fixtures having the operation or control dependent on one or more energized components, such as motors, controls, heating elements, or pressure- or temperature-sensing elements.

Such fixtures are manually adjusted or controlled by the owner or operator, or are operated automatically through one or more of the following actions: a time cycle, a temperature range, a pressure range, a measured volume or weight.

PLUMBING APPURTENANCE. A manufactured device, prefabricated assembly or an on-the-job assembly of component parts that is an adjunct to the basic piping system and plumbing fixtures. An appurtenance demands no additional water supply and does not add any discharge load to a fixture or to the drainage system.

PLUMBING FIXTURE. A receptacle or device that is either permanently or temporarily connected to the water distribution system of the premises and demands a supply of water therefrom; discharges wastewater, liquid-borne waste materials or sewage either directly or indirectly to the drainage system of the premises; or requires both a water supply connection and a discharge to the drainage system of the premises.

PLUMBING SYSTEM. Includes the water supply and distribution pipes; plumbing fixtures and traps; water-treating or water-using equipment; soil, waste and vent pipes; and sanitary and storm sewers and building drains; in addition to their respective connections, devices and appurtenances within a structure or premises.

POLLUTION. An impairment of the quality of the potable water to a degree that does not create a hazard to the public health but that does adversely and unreasonably affect the aesthetic qualities of such potable water for domestic use.

POTABLE WATER. Water free from impurities present in amounts sufficient to cause disease or harmful physiological effects and conforming to the bacteriological and chemical quality requirements of the Public Health Service Drinking Water Standards or the regulations of the public health authority having jurisdiction.

PRIVATE. In the classification of plumbing fixtures, "private" applies to fixtures in residences and apartments, and to fixtures in nonpublic toilet rooms of hotels and motels and similar installations in buildings where the plumbing fixtures are intended for utilization by a family or an individual.

PUBLIC OR PUBLIC UTILIZATION. In the classification of plumbing fixtures, "public" applies to fixtures in general toilet rooms of schools, gymnasiums, hotels, airports, bus and railroad stations, public buildings, bars, public comfort stations, office buildings, stadiums, stores, restaurants and other installations where a number of fixtures are installed so that their utilization is similarly unrestricted.

PUBLIC WATER MAIN. A water supply pipe for public utilization controlled by public authority.

QUICK-CLOSING VALVE. A valve or faucet that closes automatically when released manually or that is controlled by a mechanical means for fast-action closing.

READY ACCESS. That which enables a fixture, appliance or equipment to be directly reached without requiring the removal or movement of any panel, door or similar obstruction and without the use of a portable ladder, step stool or similar device.

REDUCED PRESSURE PRINCIPLE BACKFLOW PREVENTER. A backflow prevention device consisting of two independently acting check valves, internally force-loaded to a normally closed position and separated by an intermediate chamber (or zone) in which there is an automatic relief means of venting to the atmosphere, internally loaded to a normally open position between two tightly closing shutoff valves and with a means for testing for tightness of the checks and opening of the relief means.

REGISTERED DESIGN PROFESSIONAL. An individual who is registered or licensed to practice professional architecture or engineering as defined by the statutory requirements of the professional registration laws of the state or jurisdiction in which the project is to be constructed.

RELIEF VALVE

Pressure relief valve. A pressure-actuated valve held closed by a spring or other means and designed to relieve pressure automatically at the pressure at which such valve is set.

Temperature and pressure relief (T&P) valve. A combination relief valve designed to function as both a temperature relief and a pressure relief valve.

Temperature relief valve. A temperature-actuated valve designed to discharge automatically at the temperature at which such valve is set.

RELIEF VENT. A vent whose primary function is to provide circulation of air between drainage and vent systems.

RIM. An unobstructed open edge of a fixture.

RISER. See "Water pipe, riser."

ROOF DRAIN. A drain installed to receive water collecting on the surface of a roof and to discharge such water into a leader or a conductor.

ROUGH-IN. Parts of the plumbing system that are installed prior to the installation of fixtures. This includes drainage, water supply, vent piping and the necessary fixture supports and any fixtures that are built into the structure.

SELF-CLOSING FAUCET. A faucet containing a valve that automatically closes upon deactivation of the opening means.

SEPARATOR. See "Interceptor."

SEWAGE. Any liquid waste containing animal or vegetable matter in suspension or solution, including liquids containing chemicals in solution.

SEWAGE EJECTORS. A device for lifting sewage by entraining the sewage in a high-velocity jet of steam, air or water.

SEWER

> **Building sewer.** See "Building sewer."

> **Public sewer.** A common sewer directly controlled by public authority.

> **Sanitary sewer.** A sewer that carries sewage and excludes storm, surface and ground water.

> **Storm sewer.** A sewer that conveys rainwater, surface water, subsurface water and similar liquid wastes.

SLOPE. The fall (pitch) of a line of pipe in reference to a horizontal plane. In drainage, the slope is expressed as the fall in units vertical per units horizontal (percent) for a length of pipe.

SOIL PIPE. A pipe that conveys sewage containing fecal matter to the building drain or building sewer.

SPILLPROOF VACUUM BREAKER. An assembly consisting of one check valve force-loaded closed and an air-inlet vent valve force-loaded open to atmosphere, positioned downstream of the check valve, and located between and including two tightly closing shutoff valves and a test cock.

STACK. A general term for any vertical line of soil, waste, vent or inside conductor piping that extends through at least one story with or without offsets.

STACK VENT. The extension of a soil or waste stack above the highest horizontal drain connected to the stack.

STACK VENTING. A method of venting a fixture or fixtures through the soil or waste stack.

STERILIZER

> **Boiling type.** A boiling-type sterilizer is a fixture of a nonpressure type utilized for boiling instruments, utensils or other equipment for disinfection. These devices are portable or are connected to the plumbing system.

> **Instrument.** A device for the sterilization of various instruments.

> **Pressure (autoclave).** A pressure vessel fixture designed to utilize steam under pressure for sterilizing.

> **Pressure instrument washer sterilizer.** A pressure instrument washer sterilizer is a pressure vessel fixture designed to both wash and sterilize instruments during the operating cycle of the fixture.

> **Utensil.** A device for the sterilization of utensils as utilized in health care services.

> **Water.** A water sterilizer is a device for sterilizing water and storing sterile water.

STERILIZER VENT. A separate pipe or stack, indirectly connected to the building drainage system at the lower terminal, that receives the vapors from nonpressure sterilizers, or the exhaust vapors from pressure sterilizers, and conducts the vapors directly to the open air. Also called vapor, steam, atmospheric or exhaust vent.

STORM DRAIN. See "Drainage system, storm."

STRUCTURE. That which is built or constructed or a portion thereof.

SUBSOIL DRAIN. A drain that collects subsurface water or seepage water and conveys such water to a place of disposal.

SUMP. A tank or pit that receives sewage or liquid waste, located below the normal grade of the gravity system and that must be emptied by mechanical means.

SUMP PUMP. An automatic water pump powered by an electric motor for the removal of drainage, except raw sewage, from a sump, pit or low point.

SUMP VENT. A vent from pneumatic sewage ejectors, or similar equipment, that terminates separately to the open air.

SUPPORTS. Devices for supporting and securing pipe, fixtures and equipment.

SWIMMING POOL. Any structure, basin, chamber or tank containing an artificial body of water for swimming, diving or recreational bathing having a depth of 2 feet (610 mm) or more at any point.

TEMPERED WATER. Water having a temperature range between 85°F (29°C) and 110°F (43°C).

THIRD-PARTY CERTIFICATION AGENCY. An approved agency operating a product or material certification sytem that incorporates initial product testing, assessment and surveillance of a manufacturer's quality control system.

THIRD-PARTY CERTIFIED. Certification obtained by the manufacturer indicating that the function and performance characteristics of a product or material have been determined by testing and ongoing surveillance by an approved third-party certification agency. Assertion of certification is in the form of identification in accordance with the requirements of the third-party certification agency.

THIRD-PARTY TESTED. Procedure by which an approved testing laboratory provides documentation that a product, material or system conforms to specified requirements.

TRAP. A fitting or device that provides a liquid seal to prevent the emission of sewer gases without materially affecting the flow of sewage or wastewater through the trap.

TRAP SEAL. The vertical distance between the weir and the top of the dip of the trap.

UNSTABLE GROUND. Earth that does not provide a uniform bearing for the barrel of the sewer pipe between the joints at the bottom of the pipe trench.

VACUUM. Any pressure less than that exerted by the atmosphere.

VACUUM BREAKER. A type of backflow preventer installed on openings subject to normal atmospheric pressure that prevents backflow by admitting atmospheric pressure through ports to the discharge side of the device.

VENT PIPE. See "Vent system."

VENT STACK. A vertical vent pipe installed primarily for the purpose of providing circulation of air to and from any part of the drainage system.

VENT SYSTEM. A pipe or pipes installed to provide a flow of air to or from a drainage system, or to provide a circulation of air within such system to protect trap seals from siphonage and backpressure.

VERTICAL PIPE. Any pipe or fitting that makes an angle of 45 degrees (0.79 rad) or more with the horizontal.

WALL-HUNG WATER CLOSET. A wall-mounted water closet installed in such a way that the fixture does not touch the floor.

WASTE. The discharge from any fixture, appliance, area or appurtenance that does not contain fecal matter.

WASTE PIPE. A pipe that conveys only waste.

WATER-HAMMER ARRESTOR. A device utilized to absorb the pressure surge (water hammer) that occurs when water flow is suddenly stopped in a water supply system.

WATER HEATER. Any heating appliance or equipment that heats potable water and supplies such water to the potable hot water distribution system.

WATER MAIN. A water supply pipe or system of pipes, installed and maintained by a city, township, county, public utility company or other public entity, on public property, in the street or in an approved dedicated easement of public or community use.

WATER OUTLET. A discharge opening through which water is supplied to a fixture, into the atmosphere (except into an open tank that is part of the water supply system), to a boiler or heating system, or to any devices or equipment requiring water to operate but which are not part of the plumbing system.

WATER PIPE

　Riser. A water supply pipe that extends one full story or more to convey water to branches or to a group of fixtures.

　Water distribution pipe. A pipe within the structure or on the premises that conveys water from the water service pipe, or from the meter when the meter is at the structure, to the points of utilization.

　Water service pipe. The pipe from the water main or other source of potable water supply, or from the meter when the meter is at the public right of way, to the water distribution system of the building served.

WATER SUPPLY SYSTEM. The water service pipe, water distribution pipes, and the necessary connecting pipes, fittings, control valves and all appurtenances in or adjacent to the structure or premises.

WELL

　Bored. A well constructed by boring a hole in the ground with an auger and installing a casing.

　Drilled. A well constructed by making a hole in the ground with a drilling machine of any type and installing casing and screen.

　Driven. A well constructed by driving a pipe in the ground. The drive pipe is usually fitted with a well point and screen.

　Dug. A well constructed by excavating a large-diameter shaft and installing a casing.

WHIRLPOOL BATHTUB. A plumbing appliance consisting of a bathtub fixture that is equipped and fitted with a circulating piping system designed to accept, circulate and discharge bathtub water upon each use.

YOKE VENT. A pipe connecting upward from a soil or waste stack to a vent stack for the purpose of preventing pressure changes in the stacks.

CHAPTER 3

GENERAL REGULATIONS

SECTION 301
GENERAL

301.1 Scope. The provisions of this chapter shall govern the general regulations regarding the installation of plumbing not specific to other chapters.

301.2 System installation. Plumbing shall be installed with due regard to preservation of the strength of structural members and prevention of damage to walls and other surfaces through fixture usage.

301.3 Connections to the sanitary drainage system. All plumbing fixtures, drains, appurtenances and appliances used to receive or discharge liquid wastes or sewage shall be directly connected to the sanitary drainage system of the building or premises, in accordance with the requirements of this code. This section shall not be construed to prevent the indirect waste systems required by Chapter 8.

301.4 Connections to water supply. Every plumbing fixture, device or appliance requiring or using water for its proper operation shall be directly or indirectly connected to the water supply system in accordance with the provisions of this code.

301.5 Pipe, tube and fitting sizes. Unless otherwise specified, the pipe, tube and fitting sizes specified in this code are expressed in nominal or standard sizes as designated in the referenced material standards.

301.6 Prohibited locations. Plumbing systems shall not be located in an elevator shaft or in an elevator equipment room.

> **Exception:** Floor drains, sumps and sump pumps shall be permitted at the base of the shaft provided they are indirectly connected to the plumbing system.

301.7 Conflicts. Where conflicts between this code and the conditions of the listing or the manufacturer's installation instructions occur, the provisions of this code apply.

> **Exception:** Where a code provision is less restrictive than the conditions of the listing of the equipment or appliance or the manufacturer's installation instructions, the conditions of the listing and manufacturer's installation instructions shall apply.

SECTION 302
EXCLUSION OF MATERIALS DETRIMENTAL
TO THE SEWER SYSTEM

302.1 Detrimental or dangerous materials. Ashes, cinders or rags; flammable, poisonous or explosive liquids or gases; oil, grease or any other insoluble material capable of obstructing, damaging or overloading the building drainage or sewer system, or capable of interfering with the normal operation of the sewage treatment processes, shall not be deposited, by any means, into such systems.

302.2 Industrial wastes. Waste products from manufacturing or industrial operations shall not be introduced into the public sewer until it has been determined by the code official or other authority having jurisdiction that the introduction thereof will not damage the public sewer system or interfere with the functioning of the sewage treatment plant.

SECTION 303
MATERIALS

303.1 Identification. Each length of pipe and each pipe fitting, trap, fixture, material and device utilized in a plumbing system shall bear the identification of the manufacturer.

303.2 Installation of materials. All materials used shall be installed in strict accordance with the standards under which the materials are accepted and approved. In the absence of such installation procedures, the manufacturer's installation instructions shall be followed. Where the requirements of referenced standards or manufacturer's installation instructions do not conform to minimum provisions of this code, the provisions of this code shall apply.

303.3 Plastic pipe, fittings and components. All plastic pipe, fittings and components shall be third-party certified as conforming to NSF 14.

303.4 Third-party testing and certification. All plumbing products and materials shall comply with the referenced standards, specifications and performance criteria of this code and shall be identified in accordance with Section 303.1. When required by Table 303.4, plumbing products and materials shall either be tested by an approved third-party testing agency or certified by an approved third-party certification agency.

SECTION 304
RODENT PROOFING

304.1 General. Plumbing systems shall be designed and installed in accordance with Sections 304.2 through 304.4 to prevent rodents from entering structures.

304.2 Strainer plates. All strainer plates on drain inlets shall be designed and installed so that all openings are not greater than 0.5 inch (12.7 mm) in least dimension.

304.3 Meter boxes. Meter boxes shall be constructed in such a manner that rodents are prevented from entering a structure by way of the water service pipes connecting the meter box and the structure.

304.4 Openings for pipes. In or on structures where openings have been made in walls, floors or ceilings for the passage of pipes, such openings shall be closed and protected by the installation of approved metal collars that are securely fastened to the adjoining structure.

TABLE 303.4
PRODUCTS AND MATERIALS REQUIRING THIRD-PARTY TESTING AND THIRD-PARTY CERTIFICATION

PRODUCT OR MATERIAL	THIRD-PARTY CERTIFIED	THIRD-PARTY TESTED
Portable water supply system components and potable water fixture fittings	Required	—
Sanitary drainage and vent system components	Plastic pipe, fittings and pipe-related components	All others
Waste fixture fittings	Plastic pipe, fittings and pipe-related components	All others
Storm drainage system components	Plastic pipe, fittings and pipe-related components	All others
Plumbing fixtures	—	Required
Plumbing appliances	Required	—
Backflow prevention devices	Required	—
Water distribution system safety devices	Required	—
Special waste system components	—	Required
Subsoil drainage system components	—	Required

SECTION 305
PROTECTION OF PIPES AND
PLUMBING SYSTEM COMPONENTS

305.1 Corrosion. Pipes passing through concrete or cinder walls and floors or other corrosive material shall be protected against external corrosion by a protective sheathing or wrapping or other means that will withstand any reaction from the lime and acid of concrete, cinder or other corrosive material. Sheathing or wrapping shall allow for expansion and contraction of piping to prevent any rubbing action. Minimum wall thickness of material shall be 0.025 inch (0.64 mm).

305.2 Breakage. Pipes passing through or under walls shall be protected from breakage.

305.3 Stress and strain. Piping in a plumbing system shall be installed so as to prevent strains and stresses that exceed the structural strength of the pipe. Where necessary, provisions shall be made to protect piping from damage resulting from expansion, contraction and structural settlement.

305.4 Sleeves. Annular spaces between sleeves and pipes shall be filled or tightly caulked in an approved manner. Annular spaces between sleeves and pipes in fire-resistance-rated assemblies shall be filled or tightly caulked in accordance with the *International Building Code.*

305.5 Pipes through or under footings or foundation walls. Any pipe that passes under a footing or through a foundation wall shall be provided with a relieving arch, or a pipe sleeve pipe shall be built into the foundation wall. The sleeve shall be two pipe sizes greater than the pipe passing through the wall.

305.6 Freezing. Water, soil and waste pipes shall not be installed outside of a building, in attics or crawl spaces, concealed in outside walls, or in any other place subjected to freezing temperature unless adequate provision is made to protect such pipes from freezing by insulation or heat or both. Exterior water supply system piping shall be installed not less

than 6 inches (152 mm) below the frost line and not less than 12 inches (305 mm) below grade.

305.6.1 Sewer depth. Building sewers that connect to private sewage disposal systems shall be a minimum of [NUMBER] inches (mm) below finished grade at the point of septic tank connection. Building sewers shall be a minimum of [NUMBER] inches (mm) below grade.

305.7 Waterproofing of openings. Joints at the roof and around vent pipes, shall be made water tight by the use of lead, copper, galvanized steel, aluminum, plastic or other approved flashings or flashing material. Exterior wall openings shall be made water tight.

305.8 Protection against physical damage. In concealed locations where piping, other than cast-iron or galvanized steel, is installed through holes or notches in studs, joists, rafters or similar members less than 1.5 inches (38 mm) from the nearest edge of the member, the pipe shall be protected by shield plates. Protective shield plates shall be a minimum of 0.062-inch-thick (1.6 mm) steel, shall cover the area of the pipe where the member is notched or bored, and shall extend a minimum of 2 inches (51 mm) above sole plates and below top plates.

305.9 Protection of components of plumbing system. Components of a plumbing system installed along alleyways, driveways, parking garages or other locations exposed to damage shall be recessed into the wall or otherwise protected in an approved manner.

SECTION 306
TRENCHING, EXCAVATION AND BACKFILL

306.1 Support of piping. Buried piping shall be supported throughout its entire length.

306.2 Trenching and bedding. Where trenches are excavated such that the bottom of the trench forms the bed for the pipe, solid and continuous load-bearing support shall be provided between joints. Bell holes, hub holes and coupling holes shall be provided at points where the pipe is joined. Such pipe shall not be supported on blocks to grade. In instances where the materials manufacturer's installation instructions are more restrictive than those prescribed by the code, the material shall be installed in accordance with the more restrictive requirement.

306.2.1 Overexcavation. Where trenches are excavated below the installation level of the pipe such that the bottom of the trench does not form the bed for the pipe, the trench shall be backfilled to the installation level of the bottom of the pipe with sand or fine gravel placed in layers of 6 inches (152 mm) maximum depth and such backfill shall be compacted after each placement.

306.2.2 Rock removal. Where rock is encountered in trenching, the rock shall be removed to a minimum of 3 inches (76 mm) below the installation level of the bottom of the pipe, and the trench shall be backfilled to the installation level of the bottom of the pipe with sand tamped in place so as to provide uniform load-bearing support for the pipe between joints. The pipe, including the joints, shall not rest on rock at any point.

306.2.3 Soft load-bearing materials. If soft materials of poor load-bearing quality are found at the bottom of the trench, stabilization shall be achieved by overexcavating a minimum of two pipe diameters and backfilling to the installation level of the bottom of the pipe with fine gravel, crushed stone or a concrete foundation. The concrete foundation shall be bedded with sand tamped into place so as to provide uniform load-bearing support for the pipe between joints.

306.3 Backfilling. Backfill shall be free from discarded construction material and debris. Loose earth free from rocks, broken concrete and frozen chunks shall be placed in the trench in 6-inch (152 mm) layers and tamped in place until the crown of the pipe is covered by 12 inches (305 mm) of tamped earth. The backfill under and beside the pipe shall be compacted for pipe support. Backfill shall be brought up evenly on both sides of the pipe so that the pipe remains aligned. In instances where the manufacturer's installation instructions for materials are more restrictive than those prescribed by the code, the material shall be installed in accordance with the more restrictive requirement.

306.4 Tunneling. Where pipe is to be installed by tunneling, jacking or a combination of both, the pipe shall be protected from damage during installation and from subsequent uneven loading. Where earth tunnels are used, adequate supporting structures shall be provided to prevent future settling or caving.

SECTION 307
STRUCTURAL SAFETY

307.1 General. In the process of installing or repairing any part of a plumbing and drainage installation, the finished floors, walls, ceilings, tile work or any other part of the building or premises that must be changed or replaced shall be left in a safe structural condition in accordance with the requirements of the *International Building Code.*

307.2 Cutting, notching or bored holes. A framing member shall not be cut, notched or bored in excess of limitations specified in the *International Building Code.*

307.3 Penetrations of floor/ceiling assemblies and fire-resistance-rated assemblies. Penetrations of floor/ceiling assemblies and assemblies required to have a fire-resistance rating shall be protected in accordance with the *International Building Code.*

[B] 307.4 Alterations to trusses. Truss members and components shall not be cut, drilled, notched, spliced or otherwise altered in any way without written concurrence and approval of a registered design professional. Alterations resulting in the addition of loads to any member (e.g., HVAC equipment, water heater) shall not be permitted without verification that the truss is capable of supporting such additional loading.

307.5 Trench location. Trenches installed parallel to footings shall not extend below the 45-degree (0.79 rad) bearing plane of the footing or wall.

307.6 Piping materials exposed within plenums. All piping materials exposed within plenums shall comply with the provisions of the *International Mechanical Code.*

SECTION 308
PIPING SUPPORT

308.1 General. All plumbing piping shall be supported in accordance with this section.

308.2 Piping seismic supports. Where earthquake loads are applicable in accordance with the building code, plumbing piping supports shall be designed and installed for the seismic forces in accordance with the *International Building Code.*

308.3 Materials. Hangers, anchors and supports shall support the piping and the contents of the piping. Hangers and strapping material shall be of approved material that will not promote galvanic action.

308.4 Structural attachment. Hangers and anchors shall be attached to the building construction in an approved manner.

308.5 Interval of support. Pipe shall be supported in accordance with Table 308.5.

Exception: The interval of support for piping systems designed to provide for expansion/contraction shall conform to the engineered design in accordance with Section 105.4.

308.6 Sway bracing. Rigid support sway bracing shall be provided at changes in direction greater than 45 degrees (0.79 rad) for pipe sizes 4 inches (102 mm) and larger.

308.7 Anchorage. Anchorage shall be provided to restrain drainage piping from axial movement.

308.7.1 Location. For pipe sizes greater than 4 inches (102 mm), restraints shall be provided for drain pipes at all changes in direction and at all changes in diameter greater than two pipe sizes. Braces, blocks, rodding and other suitable methods as specified by the coupling manufacturer shall be utilized.

**TABLE 308.5
HANGER SPACING**

PIPING MATERIAL	MAXIMUM HORIZONTAL SPACING (feet)	MAXIMUM VERTICAL SPACING (feet)
ABS pipe	4	10[b]
Aluminum tubing	10	15
Brass pipe	10	10
Cast-iron pipe	5[a]	15
Copper or copper-alloy pipe	12	10
Copper or copper-allow tubing, 1^1/$_4$-inch diameter and smaller	6	10
Copper or copper-alloy tubing, 1^1/$_2$-inch diameter and larger	10	10
Cross-linked polyethylene (PEX) pipe	2.67 (32 inches)	10[b]
Cross-linked polyethylene/ Aluminum/cross-linked polyethylene (PEX-AL-PEX) pipe	2.67 (32 inches)	4[b]
CPVC pipe or tubing, 1 inch or smaller	3	10[b]
CPVC pipe or tubing, 1^1/$_4$ inches or larger	4	10[b]
Steel pipe	12	15
Lead pipe	Continuous	4
PB pipe or tubing	2.67 (32 inches)	4
Polyethylene/aluminum/polyethylene (PE-AL-PE) pipe	2.67 (32 inches)	4[b]
PVC pipe	4	10[b]
Stainless steel drainage systems	10	10[b]

For SI: 1 inch = 25.4 mm, 1 foot = 304.8 mm.

a. The maximum horizontal spacing of cast-iron pipe hangers shall be increased to 10 feet where 10-foot lengths of pipe are installed.

b. Midstory guide for sizes 2 inches and smaller.

308.8 Expansion joint fittings. Expansion joint fittings shall be used only where necessary to provide for expansion and contraction of the pipes. Expansion joint fittings shall be of the typical material suitable for use with the type of piping in which such fittings are installed.

308.9 Stacks. Bases of stacks shall be supported by concrete, brick laid in cement mortar or metal brackets attached to the building or by other approved methods.

308.10 Parallel water distribution systems. Piping bundles for manifold systems shall be supported in accordance with Table 308.5. Support at changes in direction shall be in accordance with the manufacturer's installation instructions.

Hot and cold water piping shall not be grouped in the same bundle.

SECTION 309
FLOOD HAZARD RESISTANCE

309.1 General. Plumbing systems and equipment in structures erected in flood hazard areas shall be constructed in accordance with the requirements of this section and the *International Building Code.*

[B] 309.2 Flood hazard. For structures located in flood hazard areas, the following systems and equipment shall be located at or above the design flood elevation:

Exception: The following systems are permitted to be located below the design flood elevation provided that the systems are designed and installed to prevent water from entering or accumulating within their components and the systems are constructed to resist hydrostatic and hydrodynamic loads and stresses, including the effects of buoyancy, during the occurrence of flooding to the design flood elevation.

1. All water service pipes.

2. Pump seals in individual water supply systems where the pump is located below the design flood elevation.

3. Covers on potable water wells shall be sealed, except where the top of the casing well or pipe sleeve is elevated to at least 1 foot (304.8 mm) above the design flood elevation.

4. All sanitary drainage piping.

5. All storm drainage piping.

6. Manhole covers shall be sealed, except where elevated to or above the design flood elevation.

7. All other plumbing fixtures, faucets, fixture fittings, piping systems and equipment.

8. Water heaters.

9. Vents and vent systems.

[B] 309.3 Flood hazard areas subject to high-velocity wave action. Structures located in flood hazard areas subject to high-velocity wave action shall meet the requirements of Section 309.2. The plumbing systems, pipes and fixtures shall not be mounted on or penetrate through walls intended to break away under flood loads.

SECTION 310
WASHROOM AND TOILET ROOM REQUIREMENTS

310.1 Light and ventilation. Washrooms and toilet rooms shall be illuminated and ventilated in accordance with the *International Building Code* and *International Mechanical Code.*

310.2 Location of fixtures and piping. Piping, fixtures or equipment shall not be located in such a manner as to interfere with the normal operation of windows, doors or other means of egress openings.

310.3 Interior finish. Interior finish surfaces of toilet rooms shall comply with the *International Building Code*.

310.4 Water closet compartment. Each water closet utilized by the public or employees shall occupy a separate compartment with walls or partitions and a door enclosing the fixtures to ensure privacy.

Exceptions:

1. Water closet compartments shall not be required in a single-occupant toilet room with a lockable door.

2. Toilet rooms located in day care and child-care facilities and containing two or more water closets shall be permitted to have one water closet without an enclosing compartment.

SECTION 311
TOILET FACILITIES FOR WORKERS

311.1 General. Toilet facilities shall be provided for construction workers and such facilities shall be maintained in a sanitary condition. Construction worker toilet facilities of the nonsewer type shall conform to ANSI Z4.3.

SECTION 312
TESTS AND INSPECTIONS

312.1 Required tests. The permit holder shall make the applicable tests prescribed in Sections 312.2 through 312.9 to determine compliance with the provisions of this code. The permit holder shall give reasonable advance notice to the code official when the plumbing work is ready for tests. The equipment, material, power and labor necessary for the inspection and test shall be furnished by the permit holder and the permit holder shall be responsible for determining that the work will withstand the test pressure prescribed in the following tests. All plumbing system piping shall be tested with either water or, for piping systems other than plastic, by air. After the plumbing fixtures have been set and their traps filled with water, the entire drainage system shall be submitted to final tests. The code official shall require the removal of any cleanouts if necessary to ascertain whether the pressure has reached all parts of the system.

312.1.1 Test gauges. Gauges used for testing shall be as follows:

1. Tests requiring a pressure of 10 psi or less shall utilize a testing gauge having increments of 0.10 psi or less.

2. Tests requiring a pressure of greater than 10 psi but less than or equal to 100 psi shall utilize a testing gauge having increments of 1 psi or less.

3. Tests requiring a pressure of greater than 100 psi shall utilize a testing gauge having increments of 2 psi or less.

312.2 Drainage and vent water test. A water test shall be applied to the drainage system either in its entirety or in sections. If applied to the entire system, all openings in the piping shall be tightly closed, except the highest opening, and the system shall be filled with water to the point of overflow. If the system is tested in sections, each opening shall be tightly

plugged except the highest openings of the section under test, and each section shall be filled with water, but no section shall be tested with less than a 10-foot (3048 mm) head of water. In testing successive sections, at least the upper 10 feet (3048 mm) of the next preceding section shall be tested so that no joint or pipe in the building, except the uppermost 10 feet (3048 mm) of the system, shall have been submitted to a test of less than a 10-foot (3048 mm) head of water. This pressure shall be held for at least 15 minutes. The system shall then be tight at all points.

312.3 Drainage and vent air test. An air test shall be made by forcing air into the system until there is a uniform gauge pressure of 5 pounds per square inch (psi) (34.5 kPa) or sufficient to balance a 10-inch (254 mm) column of mercury. This pressure shall be held for a test period of at least 15 minutes. Any adjustments to the test pressure required because of changes in ambient temperature or the seating of gaskets shall be made prior to the beginning of the test period.

312.4 Drainage and vent final test. The final test of the completed drainage and vent system shall be visual and in sufficient detail to determine compliance with the provisions of this code except that the plumbing shall be subjected to a smoke test where necessary for cause. Where the smoke test is utilized, it shall be made by filling all traps with water and then introducing into the entire system a pungent, thick smoke produced by one or more smoke machines. When the smoke appears at stack openings on the roof, the stack openings shall be closed and a pressure equivalent to a 1-inch water column (248.8 Pa) shall be held for a test period of not less than 15 minutes.

312.5 Water supply system test. Upon completion of a section of or the entire water supply system, the system, or portion completed, shall be tested and proved tight under a water pressure not less than the working pressure of the system; or, for piping systems other than plastic, by an air test of not less than 50 psi (344 kPa). The water utilized for tests shall be obtained from a potable source of supply. The required tests shall be performed in accordance with this section and Section 107.

312.6 Gravity sewer test. Gravity sewer tests shall consist of plugging the end of the building sewer at the point of connection with the public sewer, filling the building sewer with water, testing with not less than a 10-foot (3048 mm) head of water and maintaining such pressure for 15 minutes.

312.7 Forced sewer test. Forced sewer tests shall consist of plugging the end of the building sewer at the point of connection with the public sewer and applying a pressure of 5 psi (34.5 kPa) greater than the pump rating, and maintaining such pressure for 15 minutes.

312.8 Storm drainage system test. Storm drain systems within a building shall be tested by water or air in accordance with Section 312.2 or 312.3.

312.9 Inspection and testing of backflow prevention assemblies. Inspection and testing shall comply with Sections 312.9.1 and 312.9.2.

312.9.1 Inspections. Annual inspections shall be made of all backflow prevention assemblies and air gaps to determine whether they are operable.

312.9.2 Testing. Reduced pressure principle backflow preventer assemblies, double check-valve assemblies, pressure vacuum breaker assemblies, reduced pressure detector fire protection backflow prevention assemblies, double check detector fire protection backflow prevention assemblies, hose connection backflow preventers, and spill-proof vacuum breakers shall be tested at the time of installation, immediately after repairs or relocation and at least annually. The testing procedure shall be performed in accordance with one of the following standards:

ASSE 5013, ASSE 5015, ASSE 5020, ASSE 5047, ASSE 5048, ASSE 5052, ASSE 5056, CAN/CSA B64.10

SECTION 313
EQUIPMENT EFFICIENCIES

313.1 General. Equipment efficiencies shall be in accordance with the *International Energy Conservation Code.*

[M]SECTION 314
CONDENSATE DISPOSAL

314.1 Fuel-burning appliances. Liquid combustion byproducts of condensing appliances shall be collected and discharged to an approved plumbing fixture or disposal area in accordance with the manufacturer's installation instructions. Condensate piping shall be of approved corrosion-resistant material and shall not be smaller than the drain connection on the appliance. Such piping shall maintain a minimum horizontal slope in the direction of discharge of not less than one-eighth unit vertical in 12 units horizontal (1-percent slope).

314.2 Evaporators and cooling coils. Condensate drain systems shall be provided for equipment and appliances containing evaporators or cooling coils. Condensate drain systems shall be designed, constructed and installed in accordance with Sections 314.2.1 through 314.2.3.

314.2.1 Condensate disposal. Condensate from all cooling coils and evaporators shall be conveyed from the drain pan outlet to an approved place of disposal. Condensate shall not discharge into a street, alley or other areas so as to cause a nuisance.

314.2.2 Drain pipe materials and sizes. Components of the condensate disposal system shall be cast iron, galvanized steel, copper, cross-linked polyethylene, polybutylene, polyethylene, ABS, CPVC, or PVC pipe or tubing. All components shall be selected for the pressure and temperature rating of the installation. Condensate waste and drain line size shall not be less than $^3/_4$-inch (19 mm) internal diameter and shall not decrease in size from the drain pan connection to the place of condensate disposal. Where the drain pipes from more than one unit are manifolded together for condensate drainage, the pipe or tubing shall be sized in accordance with an approved method. All horizontal sections of drain piping shall be installed in uniform alignment at a uniform slope.

314.2.3 Auxiliary and secondary drain systems. In addition to the requirements of Section 314.2.1, a secondary drain or auxiliary drain pan shall be required for each cooling or evaporator coil where damage to any building components will occur as a result of overflow from the equipment drain pan or stoppage in the condensate drain piping. One of the following methods shall be used:

1. An auxiliary drain pan with a separate drain shall be provided under the coils on which condensation will occur. The auxiliary pan drain shall discharge to a conspicuous point of disposal to alert occupants in the event of a stoppage of the primary drain. The pan shall have a minimum depth of 1.5 inches (38 mm), shall not be less than 3 inches (76 mm) larger than the unit or the coil dimensions in width and length and shall be constructed of corrosion-resistant material. Metallic pans shall have a minimum thickness of not less than 0.0276-inch (0.7 mm) galvanized sheet metal. Nonmetallic pans shall have a minimum thickness of not less than 0.0625 inch (1.6 mm).

2. A separate overflow drain line shall be connected to the drain pan provided with the equipment. Such overflow drain shall discharge to a conspicuous point of disposal to alert occupants in the event of a stoppage of the primary drain. The overflow drain line shall connect to the drain pan at a higher level than the primary drain connection.

3. An auxiliary drain pan without a separate drain line shall be provided under the coils on which condensate will occur. Such pan shall be equipped with a water level detection device that will shut off the equipment served prior to overflow of the pan. The auxiliary drain pan shall be constructed in accordance with Item 1 of this section.

314.2.4 Traps. Condensate drains shall be trapped as required by the equipment or appliance manufacturer.

CHAPTER 4

FIXTURES, FAUCETS AND FIXTURE FITTINGS

SECTION 401
GENERAL

401.1 Scope. This chapter shall govern the materials, design and installation of plumbing fixtures, faucets and fixture fittings in accordance with the type of occupancy, and shall provide for the minimum number of fixtures for various types of occupancies.

401.2 Prohibited fixtures and connections. Water closets having a concealed trap seal or an unventilated space or having walls that are not thoroughly washed at each discharge in accordance with ASME A112.19.2M shall be prohibited. Any water closet that permits siphonage of the contents of the bowl back into the tank shall be prohibited. Trough urinals shall be prohibited.

401.3 Water conservation. The maximum water flow rates and flush volume for plumbing fixtures and fixture fittings shall comply with Section 604.4.

SECTION 402
FIXTURE MATERIALS

402.1 Quality of fixtures. Plumbing fixtures shall be constructed of approved materials, with smooth, impervious surfaces, free from defects and concealed fouling surfaces, and shall conform to standards cited in this code. All porcelain enameled surfaces on plumbing fixtures shall be acid resistant.

402.2 Materials for specialty fixtures. Materials for specialty fixtures not otherwise covered in this code shall be of stainless steel, soapstone, chemical stoneware or plastic, or shall be lined with lead, copper-base alloy, nickel-copper alloy, corrosion-resistant steel or other material especially suited to the application for which the fixture is intended.

402.3 Sheet copper. Sheet copper for general applications shall conform to ASTM B 152 and shall not weigh less than12 ounces per square foot (3.7 kg/m^2).

402.4 Sheet lead. Sheet lead for pans shall not weigh less than 4 pounds per square foot (19.5 kg/m^2) coated with an asphalt paint or other approved coating.

SECTION 403
MINIMUM PLUMBING FACILITIES

403.1 Minimum number of fixtures. Plumbing fixtures shall be provided for the type of occupancy and in the minimum number shown in Table 403.1. Types of occupancies not shown in Table 403.1 shall be considered individually by the code official. The number of occupants shall be determined by the *International Building Code*. Occupancy classification shall be determined in accordance with the *International Building Code*.

TABLE 403.1
MINIMUM NUMBER OF REQUIRED PLUMBING FIXTURES
(See Sections 403.2 and 403.3)

NO.	CLASSIFICATION	USE GROUP	DESCRIPTION	WATER CLOSETS (URINALS SEE SECTION 419.2) MALE	FEMALE	LAVATORIES MALE	FEMALE	BATHTUBS/ SHOWERS	DRINKING FOUNTAIN (SEE SECTION 410.1)	OTHER
1	Assembly (see Sections 403.2, 403.5 and 403.6)	A-1	Theaters usually with fixed seats and other buildings for the performing arts and motion pictures	1 per 125	1 per 65	1 per 200		—	1 per 500	1 service sink
		A-2	Nightclubs, bars, taverns, dance halls and buildings for similar purposes	1 per 40	1 per 40	1 per 75		—	1 per 500	1 service sink
			Restaurants, banquet halls and food courts	1 per 75	1 pe r 75	1 per 200		—	1 per 500	1 service sink
		A-3	Auditoriums without permanent seating, art galleries, exhibition halls, museums, lecture halls, libraries, arcades and gymnasiums	1 per 125	1 per 65	1 per 200		—	1 per 500	1 service sink
			Passenger terminals and transportation facilities	1 per 500	1 per 500	1 per 750		—	1 per 1,000	1 service sink
			Places of worship and other religious services. Churches without assembly halls	1 per 150	1 per 75	1 per 200		—	1 per 1,000	1 service sink

TABLE 403.1—continued
MINIMUM NUMBER OF REQUIRED PLUMBING FIXTURES
(See Sections 403.2 and 403.3)

NO.	CLASSIFICATION	USE GROUP	DESCRIPTION	WATER CLOSETS (URINALS SEE SECTION 419.2)		LAVATORIES		BATHTUBS/ SHOWERS	DRINKING FOUNTAIN (SEE SECTION 410.1)	OTHER
				MALE	FEMALE	MALE	FEMALE			
		A-4	Coliseums, arenas, skating rinks, pools and tennis courts for indoor sporting events and activities	1 per 75 for the first 1,500 and 1 per 120 for the remainder exceeding 1,500	1 per 40 for the first 1,500 and 1 per 60 for the remainer exceeding 1,500	1 per 200	1 per 150	—	1 per 1,000	1 service sink
		A-5	Stadiums, amusement parks, bleachers and grandstands for outdoor sporting events and activities	1 per 75 for the first 1,500 and 1 per 120 for the remainder exceeding 1,500	1 per 40 for the first 1,500 and 1 per 60 for the remainder exceeding 1,500	1 per 200	1 per 150	—	1 per 1,000	1 service sink
2	Business (see Sections 403.2, 403.4 and 403.6)	B	Buildings for the transaction of business, professional services, other services involving merchandise, office buildings, banks, light industrial and similar uses	1 per 25 for the first 50 and 1 per 50 for the remainder exceeding 50		1 per 40 for the first 50 and 1 per 80 for the remainder exceeding 50		—	1 per 100	1 service sink
3	Educational	E	Educational facilities	1 per 50		1 per 50		—	1 per 100	1 service sink
4	Factory and industrial	F-1 and F-2	Structures in which occupants are engaged in work fabricating, assembly or processing of products or materials	1 per 100		1 per 100		(see Section 411)	1 per 400	1 service sink
5	Institutional	I-1	Residential care	1 per 10		1 per 10		1 per 8	1 per 100	1 service sink
		I-2	Hospitals, ambulatory nursing home patients [b]	1 per room [c]		1 per room [c]		1 per 15	1 per 100	1 service sink per floor
			Employees, other than residential care [b]	1 per 25		1 per 35		—	1 per 100	—
			Visitors, other than residential care	1 per 75		1 per 100		—	1 per 500	—
		I-3	Prisons [b]	1 per cell		1 per cell		1 per 15	1 per 100	1 service sink
		I-3	Reformitories, detention centers, and correctional centers [b]	1 per 15		1 per 15		1 per 15	1 per 100	1 service sink
		I-4	Adult daycare and childcare [b]	1 per 15		1 per 15		1 per 15 [d]	1 per 100	1 service sink

(continued)

TABLE 403.1—continued
MINIMUM NUMBER OF REQUIRED PLUMBING FIXTURES
(See Sections 403.2 and 403.3)

NO.	CLASSIFICATION	USE GROUP	DESCRIPTION	WATER CLOSETS (URINALS SEE SECTION 419.2)		LAVATORIES		BATHTUBS/ SHOWERS	DRINKING FOUNTAIN (SEE SECTION 410.1)	OTHER
				MALE	FEMALE	MALE	FEMALE			
6	Mercantile (see Sections 403.2, 403.5 and 403.6)	M	Retail stores, service stations, shops, salesrooms, markets and shopping centers	1 per 500		1 per 750		—	1 per 1,000	1 service sink
7	Residential	R-1	Hotels, motels, boarding houses (transient)	1 per guestroom		1 per guestroom		1 per guestroom	—	1 service sink
		R-2	Dormitories, fraternities, sororities and boarding houses (not transient)	1 per 10		1 per 10		1 per 8	1 per 100	1 service sink
		R-2	Apartment house	1 per dwelling unit		1 per dwelling unit		1 per dwelling unit	—	1 kitchen sink per dwelling unit; 1 automatic clothes washer connection per 20 dwelling units [e]
		R-3	One- and two-family dwellings	1 per dwelling unit		1 per dwelling unit		1 per dwelling unit	—	1 kitchen sink per dwelling unit; 1 automatic clothes washer connector per dwelling unit [e]
		R-4	Residential care/assisted living facilities	1 per 10		1 per 10		1 per 8	1 per 100	1 service sink
8	Storage (see Sections 403.2 and 403.4)	S-1 S-2	Structures for the storage of goods, warehouses, storehouse and freight depots. Low and Moderate Hazard.	1 per 100		1 per 100		1 per 1,000	See Section 411	1 service sink

a. The fixtures shown are based on one fixture being the minimum required for the number of persons indicated or any fraction of the number of persons indicated. The number of occupants shall be determined by the *International Building Code*.

b. Toilet facilities for employees shall be separate from facilities for inmates or patients.

c. A single-occupant toilet room with one water closet and one lavatory serving not more than two adjacent patient rooms shall be permitted where such room is provided with direct access from each patient room and with provisions for privacy.

d. For day nurseries, a maximum of one bathtub shall be required.

e. For attached one- and two-family dwellings, one automatic clothes washer connection shall be required per 20 dwelling units.

[B]403.1.1 Unisex toilet and bath fixtures. Fixtures located within unisex toilet and bathing rooms complying with Section 404 are permitted to be included in determining the minimum required number of fixtures for assembly and mercantile occupancies.

403.2 Separate facilities. Where plumbing fixtures are required, separate facilities shall be provided for each sex.

Exceptions:

1. Separate facilities shall not be required for private facilities.

2. Separate employee facilities shall not be required in occupancies in which 15 or less people are employed.

3. Separate facilities shall not be required in structures or tenant spaces with a total occupant load, including both employees and customers, of 15 or less.

4. Separate facilities shall not be required in mercantile occupancies in which the maximum occupant load is 50 or less.

403.3 Number of occupants of each sex. The required water closets, lavatories, and showers or bathtubs shall be distributed equally between the sexes based on the percentage of each sex anticipated in the occupant load. The occupant load shall be composed of 50 percent of each sex, unless statistical data approved by the code official indicate a different distribution of the sexes.

403.4 Location of employee toilet facilities in occupancies other than assembly or mercantile. Access to toilet facilities in occupancies other than mercantile and assembly occupancies shall be from within the employees' working area. Employee facilities shall be either separate facilities or combined employee and public facilities.

> **Exception:** Facilities that are required for employees in storage structures or kiosks, and are located in adjacent structures under the same ownership, lease or control, shall be a maximum travel distance of 500 feet (152 m) from the employees' working area.

403.4.1 Travel distance. The required toilet facilities in occupancies other than assembly or mercantile shall be located not more than one story above or below the employee's working area and the path of travel to such facilities shall not exceed a distance of 500 feet (152 m).

> **Exception:** The location and maximum travel distances to required employee toilet facilities in factory and industrial occupancies are permitted to exceed that required in Section 403.4.1, provided the location and maximum travel distance are approved by the code official.

403.5 Location of employee toilet facilities in mercantile and assembly occupancies. Employees shall be provided with toilet facilities in building and tenant spaces utilized as restaurants, nightclubs, places of public assembly and mercantile occupancies. The employee facilities shall be either separate facilities or combined employee and public facilities. The required toilet facilities shall be located not more than one story above or below the employees' work area and the path of travel to such facilities, in other than covered malls, shall not exceed a distance of 500 feet (152 m). The path of travel to required facilities in covered malls shall not exceed a distance of 300 feet (91 440 mm).

> **Exception:** Employee toilet facilities shall not be required in tenant spaces where the travel distance from the main entrance of the tenant space to a central toilet area does not exceed 300 feet (91 440 mm) and such central toilet facilities are located not more than one story above or below the tenant space.

403.6 Public facilities. Customers, patrons and visitors shall be provided with public toilet facilities in structures and tenant spaces intended for public utilization. Public toilet facilities shall be located not more than one story above or below the space required to be provided with public toilet facilities and the path of travel to such facilities shall not exceed a distance of 500 feet (152 m).

403.6.1 Covered malls. In covered mall buildings, the path of travel to required toilet facilities shall not exceed a distance of 300 feet (91 440 mm). Facilities shall be installed in each individual store or in a central toilet area located in accordance with this section. The maximum travel distance to the central toilet facilities in covered mall buildings shall be measured from the main entrance of any store or tenant space.

403.6.2 Pay facilities. Where pay facilities are installed, such facilities shall be in excess of the required minimum facilities. Required facilities shall be free of charge.

403.7 Signage. Required public facilities shall be designated by a legible sign for each sex. Signs shall be readily visible and located near the entrance to each toilet facility.

SECTION 404
ACCESSIBLE PLUMBING FACILITIES

404.1 Where required. Accessible plumbing facilities and fixtures shall be provided in accordance with the *International Building Code.*

SECTION 405
INSTALLATION OF FIXTURES

405.1 Water supply protection. The supply lines and fittings for every plumbing fixture shall be installed so as to prevent backflow.

405.2 Access for cleaning. Plumbing fixtures shall be installed so as to afford easy access for cleaning both the fixture and the area around the fixture.

405.3 Setting. Fixtures shall be set level and in proper alignment with reference to adjacent walls.

405.3.1 Water closets, urinals, lavatories and bidets. A water closet, urinal, lavatory or bidet shall not be set closer than 15 inches (381 mm) from its center to any side wall, partition, vanity or other obstruction, or closer than 30 inches (762 mm) center-to-center between water closets, urinals or adjacent fixtures. There shall be at least a 21-inch (533 mm) clearance in front of the water closet, urinal or bidet to any wall, fixture or door. Water closet compartments shall not be less than 30 inches (762 mm) wide or 60 inches (1524 mm) deep. There shall be at least a 21-inch (533 mm) clearance in front of a lavatory to any wall, fixture or door (see Figure 405.3.1).

405.3.2 Public lavatories. In employee and public toilet rooms, the required lavatory shall be located in the same room as the required water closet.

405.4 Floor and wall drainage connections. Connections between the drain and floor outlet plumbing fixtures shall be made with a floor flange. The flange shall be attached to the drain and anchored to the structure. Connections between the drain and wall-hung water closets shall be made with an approved extension nipple or horn adapter. The water closet shall be bolted to the hanger with corrosion-resistant bolts or screws. Joints shall be sealed with an approved elastomeric gasket, flange-to-fixture connection complying with ASME A112.4.3 or setting compound conforming to FS TT-P-1536a.

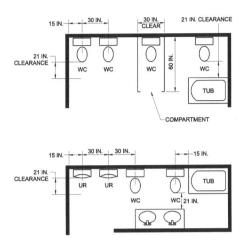

FIGURE 405.3.1
FIXTURE CLEARANCE

405.4.1 Floor flanges. Floor flanges for water closets or similar fixtures shall not be less than 0.125 inch (3.2 mm) thick for brass, 0.25 inch (6.4 mm) thick for plastic, and 0.25 inch (6.4 mm) thick and not less than a 2-inch (51 mm) caulking depth for cast-iron or galvanized malleable iron.

Floor flanges of hard lead shall weigh not less than 1 pound, 9 ounces (0.7 kg) and shall be composed of lead alloy with not less than 7.75-percent antimony by weight. Closet screws and bolts shall be of brass. Flanges shall be secured to the building structure with corrosion-resistant screws or bolts.

405.4.2 Securing floor outlet fixtures. Floor outlet fixtures shall be secured to the floor or floor flanges by screws or bolts of corrosion-resistant material.

405.4.3 Securing wall-hung water closet bowls. Wall-hung water closet bowls shall be supported by a concealed metal carrier that is attached to the building structural members so that strain is not transmitted to the closet connector or any other part of the plumbing system. The carrier shall conform to ASME A112.6.1M or ASME A112.6.2.

405.5 Water-tight joints. Joints formed where fixtures come in contact with walls or floors shall be sealed.

405.6 Plumbing in mental health centers. In mental health centers, pipes or traps shall not be exposed, and fixtures shall be bolted through walls.

405.7 Design of overflows. Where any fixture is provided with an overflow, the waste shall be designed and installed so that standing water in the fixture will not rise in the overflow when the stopper is closed, and no water will remain in the overflow when the fixture is empty.

405.7.1 Connection of overflows. The overflow from any fixture shall discharge into the drainage system on the inlet or fixture side of the trap.

Exception: The overflow from a flush tank serving a water closet or urinal shall discharge into the fixture served.

405.8 Slip joint connections. Slip joints shall be made with an approved elastomeric gasket and shall only be installed on the trap outlet, trap inlet and within the trap seal. Fixtures with concealed slip-joint connections shall be provided with an access panel or utility space at least 12 inches (305 mm) in its smallest dimension or other approved arrangement so as to provide access to the slip joint connections for inspection and repair.

405.9 Design and installation of plumbing fixtures. Integral fixture fitting mounting surfaces on manufactured plumbing fixtures or plumbing fixtures constructed on site, shall meet the design requirements of ASME A112.19.2M or ASME A112.19.3M.

SECTION 406
AUTOMATIC CLOTHES WASHERS

406.1 Approval. Domestic automatic clothes washers shall conform to ASSE 1007.

406.2 Water connection. The water supply to an automatic clothes washer shall be protected against backflow by an air gap installed integrally within the machine conforming to ASSE 1007 or with the installation of a backflow preventer in accordance with Section 608.

406.3 Waste connection. The waste from an automatic clothes washer shall discharge through an air break into a standpipe in accordance with Section 802.4 or into a laundry sink. The trap and fixture drain for an automatic clothes washer standpipe shall be a minimum of 2 inches (51 mm) in diameter. The automatic clothes washer fixture drain shall connect to a branch drain or drainage stack a minimum of 3 inches (76 mm) in diameter.

SECTION 407
BATHTUBS

407.1 Approval. Bathtubs shall conform to ANSI Z124.1, ASME A112.19.1M, ASME A112.19.4M, ASME A112.19.9M, CSA B45.2, CSA B45.3 or CSA B45.5.

407.2 Bathtub waste outlets. Bathtubs shall have waste outlets a minimum of 1.5 inches (38 mm) in diameter. The waste outlet shall be equipped with an approved stopper.

407.3 Glazing. Windows and doors within a bathtub enclosure shall conform to the safety glazing requirements of the *International Building Code.*

407.4 Bathtub enclosure. Doors within a bathtub enclosure shall conform to ASME A112.19.15.

SECTION 408
BIDETS

408.1 Approval. Bidets shall conform to ASME A112.19.2M, ASME A112.19.9M or CSA B45.1.

408.2 Water connection. The water supply to a bidet shall be protected against backflow by an air gap or backflow preventer in accordance with Section 608.13.1, 608.13.2, 608.13.3, 608.13.5, 608.13.6 or 608.13.8.

SECTION 409
DISHWASHING MACHINES

409.1 Approval. Domestic dishwashing machines shall conform to ASSE 1006. Commercial dishwashing machines shall conform to ASSE 1004 and NSF 3.

409.2 Water connection. The water supply to a dishwashing machine shall be protected against backflow by an air gap or backflow preventer in accordance with Section 608.

409.3 Waste connection. The waste connection of a dishwashing machine shall comply with Section 802.1.6 or 802.1.7, as applicable.

SECTION 410
DRINKING FOUNTAINS

410.1 Approval. Drinking fountains shall conform to ASME A112.19.1M, ASME A112.19.2M or ASME A112.19.9M, and water coolers shall conform to ARI 1010. Drinking fountains and water coolers shall conform to NSF 61, Section 9. Where water is served in restaurants, drinking fountains shall not be required. In other occupancies, where drinking fountains are required, bottled water dispensers shall be permitted to be substituted for not more than 50 percent of the required drinking fountains.

410.2 Prohibited location. Drinking fountains shall not be installed in public restrooms.

SECTION 411
EMERGENCY SHOWERS AND
EYEWASH STATIONS

411.1 Approval. Emergency showers and eyewash stations shall conform to ISEA Z358.1.

411.2 Waste connection. Waste connections shall not be required for emergency showers and eyewash stations.

SECTION 412
FLOOR AND TRENCH DRAINS

412.1 Approval. Floor drains shall conform to ASME A112.6.3, ASME A112.3.1 or CSA B79. Trench drains shall comply with ASME A112.6.3.

412.2 Floor drain trap and strainer. Floor drain traps shall have removable strainers. The strainer shall have a waterway area of not less than the area of the tailpiece. The floor drain shall be constructed so that the drain is capable of being cleaned. Access shall be provided to the drain inlet.

412.3 Size of floor drains. Floor drains shall have a minimum 2-inch-diameter (51 mm) drain outlet.

412.4 Public laundries and central washing facilities. In public coin-operated laundries and in the central washing facilities of multiple-family dwellings, the rooms containing automatic clothes washers shall be provided with floor drains located to readily drain the entire floor area. Such drains shall have a minimum outlet of not less than 3 inches (76 mm) in diameter.

SECTION 413
FOOD WASTE GRINDER UNITS

413.1 Approval. Domestic food waste grinders shall conform to ASSE 1008. Commercial food waste grinders shall conform to ASSE 1009. Food waste grinders shall not increase the drainage fixture unit load on the sanitary drainage system.

413.2 Domestic food waste grinder waste outlets. Domestic food waste grinders shall be connected to a drain of not less than 1.5 inches (38 mm) in diameter.

413.3 Commercial food waste grinder waste outlets. Commercial food waste grinders shall be connected to a drain a minimum of 2 inches (51 mm) in diameter. Commercial food waste grinders shall be connected and trapped separately from any other fixtures or sink compartments.

413.4 Water supply required. All food waste grinders shall be provided with a supply of cold water.

SECTION 414
GARBAGE CAN WASHERS

414.1 Water connection. The water supply to a garbage can washer shall be protected against backflow by an air gap or a backflow preventer in accordance with Section 608.13.1, 608.13.2, 608.13.3, 608.13.5, 608.13.6 or 608.13.8.

414.2 Waste connection. Garbage can washers shall be trapped separately. The receptacle receiving the waste from the washer shall have a removable basket or strainer to prevent the discharge of large particles into the drainage system.

SECTION 415
LAUNDRY TRAYS

415.1 Approval. Laundry trays shall conform to ANSI Z124.6, ASME A112.19.1M, ASME A112.19.3M, ASME A112.19.9M, CSA B45.2 or CSA B45.4.

415.2 Waste outlet. Each compartment of a laundry tray shall be provided with a waste outlet a minimum of 1.5 inches (38 mm) in diameter and a strainer or crossbar to restrict the clear opening of the waste outlet.

SECTION 416
LAVATORIES

416.1 Approval. Lavatories shall conform to ANSI Z124.3, ASME A112.19.1M, ASME A112.19.2M, ASME A112.19.3M, ASME A112.19.4M, ASME A112.19.9M, CSA B45.1, CSA B45.2, CSA B45.3 or CSA B45.4. Group wash-up equipment shall conform to the requirements of Section 402. Every 20 inches (508 mm) of rim space shall be considered as one lavatory.

416.2 Cultured marble lavatories. Cultured marble vanity tops with an integral lavatory shall conform to ANSI Z124.3 or CSA B45.5.

416.3 Lavatory waste outlets. Lavatories shall have waste outlets not less than 1.25 inches (32 mm) in diameter. A strainer, pop-up stopper, crossbar or other device shall be provided to restrict the clear opening of the waste outlet.

416.4 Moveable lavatory systems. Moveable lavatory systems shall comply with ASME A112.19.12.

SECTION 417
SHOWERS

417.1 Approval. Prefabricated showers and shower compartments shall conform to ANSI Z124.2, ASME A112.19.9M or CSA B45.5. Shower valves for individual showers shall conform to the requirements of Section 424.4.

417.2 Water supply riser. Every water supply riser from the shower valve to the shower head outlet, whether exposed or not, shall be attached to the structure in an approved manner.

417.3 Shower waste outlet. Waste outlets serving showers shall be at least $1^1/_2$ inches (38 mm) in diameter and, for other than waste outlets in bathtubs, shall have removable strainers not less than 3 inches (76 mm) in diameter with strainer openings not less than 0.25 inch (6.4 mm) in minimum dimension. Where each shower space is not provided with an individual waste outlet, the waste outlet shall be located and the floor pitched so that waste from one shower does not flow over the floor area serving another shower. Waste outlets shall be fastened to the waste pipe in an approved manner.

417.4 Shower compartments. All shower compartments shall have a minimum of 900 square inches (0.58 m²) of interior cross-sectional area. Shower compartments shall not be less than 30 inches (762 mm) in minimum dimension measured from the finished interior dimension of the compartment, exclusive of fixture valves, showerheads, soap dishes, and safety grab bars or rails. Except as required in Section 404, the minimum required area and dimension shall be measured from the finished interior dimension at a height equal to the top of the threshold and at a point tangent to its centerline and shall be continued to a height not less than 70 inches (1778 mm) above the shower drain outlet.

417.4.1 Wall area. The wall area above built-in tubs with installed shower heads and in shower compartments shall be constructed of smooth, noncorrosive and nonabsorbent waterproof materials to a height not less than 6 feet (1829 mm) above the room floor level, and not less than 70 inches (1778 mm) where measured from the compartment floor at the drain. Such walls shall form a water-tight joint with each other and with either the tub, receptor or shower floor.

417.5 Shower floors or receptors. Floor surfaces shall be constructed of impervious, noncorrosive, nonabsorbent and waterproof materials.

417.5.1 Support. Floors or receptors under shower compartments shall be laid on, and supported by, a smooth and structurally sound base.

417.5.2 Shower lining. Floors under shower compartments, except where prefabricated receptors have been provided, shall be lined and made water tight utilizing material complying with Sections 417.5.2.1 through 417.5.2.4. Such liners shall turn up on all sides at least 2 inches (51 mm) above the finished threshold level. Liners shall be recessed and fastened to an approved backing so as not to occupy the space required for wall covering, and shall not be nailed or perforated at any point less than 1 inch (25.4 mm) above the finished threshold. Liners shall be pitched one-fourth unit vertical in 12 units horizontal (2-percent slope) and shall be sloped toward the fixture drains and be securely fastened to the waste outlet at the seepage entrance, making a water-tight joint between the liner and the outlet.

Exception: Floor surfaces under shower heads provided for rinsing laid directly on the ground are not required to comply with this section.

417.5.2.1 PVC sheets. Plasticized polyvinyl chloride (PVC) sheets shall be a minimum of 0.040 inch (1.02 mm) thick, and shall meet the requirements of ASTM D 4551. Sheets shall be joined by solvent welding in accordance with the manufacturer's installation instructions.

417.5.2.2 Chlorinated polyethylene (CPE) sheets. Nonplasticized chlorinated polyethylene sheet shall be a minimum 0.040 inch (1.02 mm) thick, and shall meet the requirements of ASTM D 4068. The liner shall be joined in accordance with the manufacturer's installation instructions.

417.5.2.3 Sheet lead. Sheet lead shall not weigh less than 4 pounds per square foot (19.5 kg/m²) coated with an asphalt paint or other approved coating. The lead sheet shall be insulated from conducting substances other than the connecting drain by 15-pound (6.80 kg) asphalt felt or its equivalent. Sheet lead shall be joined by burning.

417.5.2.4 Sheet copper. Sheet copper shall conform to ASTM B 152 and shall not weigh less than 12 ounces per square foot (3.7 kg/m²). The copper sheet shall be insulated from conducting substances other than the connecting drain by 15-pound (6.80 kg) asphalt felt or its equivalent. Sheet copper shall be joined by brazing or soldering.

417.6 Glazing. Windows and doors within a shower enclosure shall conform to the safety glazing requirements of the *International Building Code*.

SECTION 418
SINKS

418.1 Approval. Sinks shall conform to ANSI Z124.6, ASME A112.19.1M, ASME A112.19.2M, ASME A112.19.3M, ASME A112.19.4M, ASME A112.19.9M, CSA B45.1, CSA B45.2, CSA B45.3 or CSA B45.4.

418.2 Sink waste outlets. Sinks shall be provided with waste outlets a minimum of 1.5 inches (38 mm) in diameter. A strainer or crossbar shall be provided to restrict the clear opening of the waste outlet.

418.3 Moveable sink systems. Moveable sink systems shall comply with ASME A112.19.12.

SECTION 419
URINALS

419.1 Approval. Urinals shall conform to ASME A112.19.2M, CSA B45.1 or CSA B45.5. Urinals shall conform to the water consumption requirements of Section 604.4. Urinals shall conform to the hydraulic performance requirements of ASME A112.19.6, CSA B45.1 or CSA B45.5.

419.2 Substitution for water closets. In each bathroom or toilet room, urinals shall not be substituted for more than 67 percent of the required water closets.

[B] 419.3 Surrounding material. Wall and floor space to a point 2 feet (610 mm) in front of a urinal lip and 4 feet (1219 mm) above the floor and at least 2 feet (610 mm) to each side of the urinal shall be waterproofed with a smooth, readily cleanable, nonabsorbent material.

SECTION 420
WATER CLOSETS

420.1 Approval. Water closets shall conform to the water consumption requirements of Section 604.4 and shall conform to ANSI Z124.4, ASME A112.19.2M, CSA B45.1, CSA B45.4 or CSA B45.5. Water closets shall conform to the hydraulic performance requirements of ASME A112.19.6. Water closet tanks shall conform to ANSI Z124.4, ASME A112.19.2, ASME A112.19.9M, CSA B45.1, CSA B45.4 or CSA B45.5. Electro-hydraulic water closets shall comply with ASME A112.19.13.

420.2 Water closets for public or employee toilet facilities. Water closet bowls for public or employee toilet facilities shall be of the elongated type.

420.3 Water closet seats. Water closets shall be equipped with seats of smooth, nonabsorbent material. All seats of water closets provided for public or employee toilet facilities shall be of the hinged open-front type. Integral water closet seats shall be of the same material as the fixture. Water closet seats shall be sized for the water closet bowl type.

420.4 Water closet connections. A 4-inch by 3-inch (102 mm by 76 mm) closet bend shall be acceptable. Where a 3-inch (76 mm) bend is utilized on water closets, a 4-inch by 3-inch (102 mm by 76 mm) flange shall be installed to receive the fixture horn.

SECTION 421
WHIRLPOOL BATHTUBS

421.1 Approval. Whirlpool bathtubs shall comply with ASME A112.19.7M or with CSA B45.5 and CSA B45 (Supplement 1).

421.2 Installation. Whirlpool bathtubs shall be installed and tested in accordance with the manufacturer's installation instructions. The pump shall be located above the weir of the fixture trap. Access shall be provided to the pump.

421.3 Drain. The pump drain and circulation piping shall be sloped to drain the water in the volute and the circulation piping when the whirlpool bathtub is empty.

421.4 Suction fittings. Suction fittings for whirlpool bathtubs shall comply with ASME A112.19.8M.

421.5 Whirlpool enclosure. Doors within a whirlpool enclosure shall conform to ASME A112.19.15.

SECTION 422
HEALTH CARE FIXTURES AND EQUIPMENT

422.1 Scope. This section shall govern those aspects of health care plumbing systems that differ from plumbing systems in other structures. Health care plumbing systems shall conform to the requirements of this section in addition to the other requirements of this code. The provisions of this section shall apply to the special devices and equipment installed and maintained in the following occupancies: nursing homes, homes for the aged, orphanages, infirmaries, first aid stations, psychiatric facilities, clinics, professional offices of dentists and doctors, mortuaries, educational facilities, surgery, dentistry, research and testing laboratories, establishments manufacturing pharmaceutical drugs and medicines, and other structures with similar apparatus and equipment classified as plumbing.

422.2 Approval. All special plumbing fixtures, equipment, devices and apparatus shall be of an approved type.

422.3 Protection. All devices, appurtenances, appliances and apparatus intended to serve some special function, such as sterilization, distillation, processing, cooling, or storage of ice or foods, and that connect to either the water supply or drainage system, shall be provided with protection against backflow, flooding, fouling, contamination of the water supply system and stoppage of the drain.

422.4 Materials. Fixtures designed for therapy, special cleansing or disposal of waste materials, combinations of such purposes, or any other special purpose, shall be of smooth, impervious, corrosion-resistant materials and, where subjected to temperatures in excess of 180°F (82°C), shall be capable of withstanding, without damage, higher temperatures.

422.5 Access. Access shall be provided to concealed piping in connection with special fixtures where such piping contains steam traps, valves, relief valves, check valves, vacuum breakers or other similar items that require periodic inspection, servicing, maintenance or repair. Access shall be provided to concealed piping that requires periodic inspection, maintenance or repair.

422.6 Clinical sink. A clinical sink shall have an integral trap in which the upper portion of a visible trap seal provides a water surface. The fixture shall be designed so as to permit complete removal of the contents by siphonic or blowout action and to reseal the trap. A flushing rim shall provide water to cleanse the interior surface. The fixture shall have the flushing and cleansing characteristics of a water closet.

422.7 Prohibited usage of clinical sinks and service sinks. A clinical sink serving a soiled utility room shall not be considered as a substitute for, or be utilized as, a service sink. A service sink shall not be utilized for the disposal of urine, fecal matter or other human waste.

422.8 Ice prohibited in soiled utility room. Machines for manufacturing ice, or any device for the handling or storage of ice, shall not be located in a soiled utility room.

422.9 Sterilizer equipment requirements. The approval and installation of all sterilizers shall conform to the requirements of the *International Mechanical Code*.

422.9.1 Sterilizer piping. Access for the purposes of inspection and maintenance shall be provided to all sterilizer piping and devices necessary for the operation of sterilizers.

422.9.2 Steam supply. Steam supplies to sterilizers, including those connected by pipes from overhead mains or branches, shall be drained to prevent any moisture from reaching the sterilizer. The condensate drainage from the steam supply shall be discharged by gravity.

422.9.3 Steam condensate return. Steam condensate returns from sterilizers shall be a gravity return system.

422.9.4 Condensers. Pressure sterilizers shall be equipped with a means of condensing and cooling the exhaust steam vapors. Nonpressure sterilizers shall be equipped with a device that will automatically control the vapor, confining the vapors within the vessel.

422.10 Special elevations. Control valves, vacuum outlets and devices protruding from a wall of an operating, emergency, recovery, examining or delivery room, or in a corridor or other location where patients are transported on a wheeled stretcher, shall be located at an elevation that prevents bumping the patient or stretcher against the device.

SECTION 423
SPECIALTY PLUMBING FIXTURES

423.1 Water connections. Baptisteries, ornamental and lily pools, aquariums, ornamental fountain basins, swimming pools, and similar constructions, where provided with water supplies, shall be protected against backflow in accordance with Section 608.

423.2 Approval. Specialties requiring water and waste connections shall be submitted for approval.

SECTION 424
FAUCETS AND OTHER FIXTURE FITTINGS

424.1 Approval. Faucets and fixture fittings shall conform to ASME A112.18.1 or CSA B125. Faucets and fixture fittings that supply drinking water for human ingestion shall conform to the requirements of NSF 61, Section 9. Flexible water connectors exposed to continuous pressure shall conform to the requirements of Section 605.6.

424.1.1 Faucets and supply fittings. Faucets and supply fittings shall conform to the water consumption requirements of Section 604.4.

424.1.2 Waste fittings. Waste fittings shall conform to one of the standards listed in Tables 702.1 and 702.4 for above-ground drainage and vent pipe and fittings, or the waste fittings shall be constructed of tubular stainless steel with a minimum wall thickness of 0.012 inch (0.30 mm), tubular copper alloy having a minimum wall thickness of 0.027 inch (0.69 mm) or tubular plastic complying with ASTM F 409.

424.2 Hand showers. Hand-held showers shall conform to ASSE 1014 or CSA B125.

424.3 Shower valves. Shower and tub-shower combination valves shall be balanced pressure, thermostatic or combination balanced-pressure/thermostatic valves that conform to the requirements of ASSE 1016 or CSA B125. Multiple (gang) showers supplied with a single tempered water supply pipe shall have the water supply for such showers controlled by a master thermostatic mixing valve complying with ASSE 1017. Shower and tub-shower combination valves and master thermostatic mixing valves required by this section shall be equipped with a means to limit the maximum setting of the valve to 120°F (49°C), which shall be field adjusted in accordance with the manufacturer's instructions.

424.4 Hose-connected outlets. Faucets and fixture fittings with hose-connected outlets shall conform to ASME A112.18.3M.

424.5 Temperature-actuated, flow reduction valves for individual fixture fittings. Temperature-actuated, flow reduction devices, where installed for individual fixture fittings, shall conform to ASSE 1062. Such valves shall not be used alone as a substitute for the balanced pressure, thermostatic or combination shower valves required in Section 424.3.

424.6 Transfer valves. Deck-mounted bath/shower transfer valves containing an integral atmospheric vacuum breaker shall conform to the requirements of ASME A112.18.7.

SECTION 425
FLUSHING DEVICES FOR WATER CLOSETS
AND URINALS

425.1 Flushing devices required. Each water closet, urinal, clinical sink and any plumbing fixture that depends on trap siphonage to discharge the fixture contents to the drainage system shall be provided with a flushometer valve, flushometer tank or a flush tank designed and installed to supply water in quantity and rate of flow to flush the contents of the fixture, cleanse the fixture and refill the fixture trap.

425.1.1 Separate for each fixture. A flushing device shall not serve more than one fixture.

425.2 Flushometer valves and tanks. Flushometer valves and tanks shall comply with ASSE 1037. Vacuum breakers on flushometer valves shall conform to the performance requirements of ASSE 1001 or CAN/CSA-B64.1.1. Access shall be provided to vacuum breakers. Flushometer valves shall be of the water-conservation type and shall not be utilized where the water pressure is lower than the minimum required for normal operation. When operated, the valve shall automatically complete the cycle of operation, opening fully and closing positively under the water supply pressure. Each flushometer valve shall be provided with a means for regulating the flow through the valve. The trap seal to the fixture shall be automatically refilled after each valve flushing cycle.

425.3 Flush tanks. Flush tanks equipped for manual flushing shall be controlled by a device designed to refill the tank after each discharge and to shut off completely the water flow to the tank when the tank is filled to operational capacity. The trap seal to the fixture shall be automatically refilled after each flushing. The water supply to flush tanks equipped for automatic flushing shall be controlled with a timing device or sensor control devices.

425.3.1 Fill valves. All flush tanks shall be equipped with an antisiphon fill valve conforming to ASSE 1002 or CSA B125. The fill valve backflow preventer shall be located at least 1 inch (25 mm) above the full opening of the overflow pipe.

425.3.2 Overflows in flush tanks. Flush tanks shall be provided with overflows discharging to the water closet or urinal connected thereto and shall be sized to prevent flooding the tank at the maximum rate at which the tanks are supplied with water according to the manufacturer's design conditions. The opening of the overflow pipe shall be located above the flood level rim of the water closet or urinal or above a secondary overflow in the flush tank.

425.3.3 Sheet copper. Sheet copper utilized for flush tank linings shall conform to ASTM B 152 and shall not weigh less than 10 ounces per square foot (0.03 kg/m^2).

425.3.4 Access required. All parts in a flush tank shall be accessible for repair and replacement.

425.5 Flush pipes and fittings. Flush pipes and fittings shall be of nonferrous material and shall conform to ASME A112.19.5 or CSA B125.

SECTION 426
MANUAL FOOD AND BEVERAGE
DISPENSING EQUIPMENT

426.1 Approval. Manual food and beverage dispensing equipment shall conform to the requirements of NSF 18.

SECTION 427
FLOOR SINKS

427.1 Approval. Sanitary floor sinks shall conform to the requirements of ASME A112.6.7.

CHAPTER 5
WATER HEATERS

SECTION 501
GENERAL

501.1 Scope. The provisions of this chapter shall govern the materials, design and installation of water heaters and the related safety devices and appurtenances.

501.2 Water heater as space heater. Where a combination potable water heating and space heating system requires water for space heating at temperatures higher than 140°F (60°C), a master thermostatic mixing valve complying with ASSE 1017 shall be provided to limit the water supplied to the potable hot water distribution system to a temperature of 140°F (60°C) or less. The potability of the water shall be maintained throughout the system.

501.3 Drain valves. Drain valves for emptying shall be installed at the bottom of each tank-type water heater and hot water storage tank. Drain valves shall conform to ASSE 1005.

501.4 Location. Water heaters and storage tanks shall be located and connected so as to provide access for observation, maintenance, servicing and replacement.

501.5 Water heater labeling. All water heaters shall be third-party certified.

501.6 Water temperature control in piping from tankless heaters. The temperature of water from tankless water heaters shall be a maximum of 140°F (60°C) when intended for domestic uses. This provision shall not supersede the requirement for protective shower valves in accordance with Section 424.3.

501.7 Pressure marking of storage tanks. Storage tanks and water heaters installed for domestic hot water shall have the maximum allowable working pressure clearly and indelibly stamped in the metal or marked on a plate welded thereto or otherwise permanently attached. Such markings shall be in an accessible position outside of the tank so as to make inspection or reinspection readily possible.

501.8 Temperature controls. All hot water supply systems shall be equipped with automatic temperature controls capable of adjustments from the lowest to the highest acceptable temperature settings for the intended temperature operating range.

SECTION 502
INSTALLATION

502.1 General. Water heaters shall be installed in accordance with the manufacturer's installation instructions. Oil-fired water heaters shall conform to the requirements of this code and the *International Mechanical Code.* Electric water heaters shall conform to the requirements of this code and provisions of the ICC *Electrical Code* listed in Chapter 13. Gas-fired water heaters shall conform to the requirements of the *International Fuel Gas Code.*

502.2 Rooms used as a plenum. Water heaters using solid, liquid or gas fuel shall not be installed in a room containing air-handling machinery when such room is used as a plenum.

502.3 Water heaters installed in attics. Attics containing a water heater shall be provided with an opening and unobstructed passageway large enough to allow removal of the water heater. The passageway shall not be less than 30 inches (762 mm) high and 22 inches (559 mm) wide and not more than 20 feet (6096 mm) in length when measured along the centerline of the passageway from the opening to the water heater. The passageway shall have continuous solid flooring not less than 24 inches (610 mm) wide. A level service space at least 30 inches (762 mm) deep and 30 inches (762 mm) wide shall be present at the front or service side of the water heater. The clear access opening dimensions shall be a minimum of 20 inches by 30 inches (508 mm by 762 mm) where such dimensions are large enough to allow removal of the water heater.

502.4 Seismic supports. Where earthquake loads are applicable in accordance with the *International Building Code,* water heater supports shall be designed and installed for the seismic forces in accordance with the *International Building Code.*

SECTION 503
CONNECTIONS

503.1 Cold water line valve. The cold water branch line from the main water supply line to each hot water storage tank or water heater shall be provided with a valve, located near the equipment and serving only the hot water storage tank or water heater. The valve shall not interfere or cause a disruption of the cold water supply to the remainder of the cold water system. The valve shall be provided with access on the same floor level as the water heater served.

503.2 Water circulation. The method of connecting a circulating water heater to the tank shall provide proper circulation of water through the water heater. The pipe or tubes required for the installation of appliances that will draw from the water heater or storage tank shall comply with the provisions of this code for material and installation.

SECTION 504
SAFETY DEVICES

504.1 Antisiphon devices. An approved means, such as a cold water "dip" tube with a hole at the top or a vacuum relief valve installed in the cold water supply line above the top of the heater or tank, shall be provided to prevent siphoning of any storage water heater or tank.

504.2 Vacuum relief valve. Bottom fed water heaters and bottom fed tanks connected to water heaters shall have a vacuum relief valve installed. The vacuum relief valve shall comply with ANSI Z21.22.

504.3 Shutdown. A means for disconnecting an electric hot water supply system from its energy supply shall be provided in accordance with the *ICC Electrical Code.* A separate valve shall be provided to shut off the energy fuel supply to all other types of hot water supply systems.

504.4 Relief valve. All storage water heaters operating above atmospheric pressure shall be provided with an approved, self-closing (levered) pressure relief valve and temperature relief valve or combination thereof. The relief valve shall conform to ANSI Z21.22. The relief valve shall not be used as a means of controlling thermal expansion.

504.4.1 Installation. Such valves shall be installed in the shell of the water heater tank. Temperature relief valves shall be so located in the tank as to be actuated by the water in the top 6 inches (152 mm) of the tank served. For installations with separate storage tanks, the valves shall be installed on the tank and there shall not be any type of valve installed between the water heater and the storage tank. There shall not be a check valve or shutoff valve between a relief valve and the heater or tank served.

504.5 Relief valve approval. Temperature and pressure relief valves, or combinations thereof, and energy cutoff devices shall bear the label of an approved agency and shall have a temperature setting of not more than 210°F (99°C) and a pressure setting not exceeding the tank or water heater manufacturer's rated working pressure or 150 psi (1035 kPa), whichever is less. The relieving capacity of each pressure relief valve and each temperature relief valve shall equal or exceed the heat input to the water heater or storage tank.

504.6 Relief outlet waste. The outlet of a pressure, temperature or other relief valve shall not be directly connected to the drainage system.

504.6.1 Discharge. The relief valve shall discharge full size to a safe place of disposal such as the floor, outside the building, or an indirect waste receptor. The discharge pipe shall not have any trapped sections and shall have a visible air gap or air gap fitting located in the same room as the water heater. The outlet end of the discharge pipe shall not be threaded and such discharge pipe shall not have a valve or tee installed. Relief valve piping shall be piped independent of other equipment drains or relief valve discharge piping to the disposal point. Such pipe shall be installed in a manner that does not cause personal injury to occupants in the immediate area or structural damage to the building.

504.6.2 Materials. Relief valve discharge piping shall be of those materials listed in Section 605.4 or shall be tested, rated and approved for such use in accordance with ASME A112.4.1. Piping from safety pan drains shall be of those materials listed in Table 605.4.

504.7 Required pan. Where water heaters or hot water storage tanks are installed in locations where leakage of the tanks or connections will cause damage, the tank or water heater shall be installed in a galvanized steel pan having a minimum thickness of 24 gage, or other pans approved for such use.

504.7.1 Pan size and drain. The pan shall be not less than 1.5 inches (38 mm) deep and shall be of sufficient size and shape to receive all dripping or condensate from the tank or water heater. The pan shall be drained by an indirect waste pipe having a minimum diameter of $^3/_4$ inch (19 mm).

504.7.2 Pan drain termination. The pan drain shall extend full-size and terminate over a suitably located indirect waste receptor or floor drain or extend to the exterior of the building and terminate not less than 6 inches (152 mm) and not more than 24 inches (610 mm) above the adjacent ground surface.

SECTION 505
INSULATION

[E] 505.1 Unfired vessel insulation. Unfired hot water storage tanks shall be insulated so that heat loss is limited to a maximum of 15 British thermal units per hour (Btu/h) per square foot (47 W/m²) of external tank surface area. For purposes of determining this heat loss, the design ambient temperature shall not be higher than 65°F (18°C).

CHAPTER 6

WATER SUPPLY AND DISTRIBUTION

SECTION 601
GENERAL

601.1 Scope. This chapter shall govern the materials, design and installation of water supply systems, both hot and cold, for utilization in connection with human occupancy and habitation and shall govern the installation of individual water supply systems.

601.2 Solar energy utilization. Solar energy systems used for heating potable water or using an independent medium for heating potable water shall comply with the applicable requirements of this code. The use of solar energy shall not compromise the requirements for cross connection or protection of the potable water supply system required by this code.

601.3 Existing piping used for grounding. Existing metallic water service piping used for electrical grounding shall not be replaced with nonmetallic pipe or tubing until other approved means of grounding is provided.

601.4 Tests. The potable water distribution system shall be tested in accordance with Section 312.5.

SECTION 602
WATER REQUIRED

602.1 General. Every structure equipped with plumbing fixtures and utilized for human occupancy or habitation shall be provided with a potable supply of water in the amounts and at the pressures specified in this chapter.

602.2 Potable water required. Only potable water shall be supplied to plumbing fixtures that provide water for drinking, bathing or culinary purposes, or for the processing of food, medical or pharmaceutical products. Unless otherwise provided in this code, potable water shall be supplied to all plumbing fixtures.

602.3 Individual water supply. Where a potable public water supply is not available, individual sources of potable water supply shall be utilized.

602.3.1 Sources. Dependent on geological and soil conditions and the amount of rainfall, individual water supplies are of the following types: drilled well, driven well, dug well, bored well, spring, stream or cistern. Surface bodies of water and land cisterns shall not be sources of individual water supply unless properly treated by approved means to prevent contamination.

602.3.2 Minimum quantity. The combined capacity of the source and storage in an individual water supply system shall supply the fixtures with water at rates and pressures as required by this chapter.

602.3.3 Water quality. Water from an individual water supply shall be approved as potable by the authority having jurisdiction prior to connection to the plumbing system.

602.3.4 Disinfection of system. After construction or major repair, the individual water supply system shall be purged of deleterious matter and disinfected in accordance with Section 610.

602.3.5 Pumps. Pumps shall be rated for the transport of potable water. Pumps in an individual water supply system shall be constructed and installed so as to prevent contamination from entering a potable water supply through the pump units. Pumps shall be sealed to the well casing or covered with a water-tight seal. Pumps shall be designed to maintain a prime and installed such that ready access is provided to the pump parts of the entire assembly for repairs.

602.3.5.1 Pump enclosure. The pump room or enclosure around a well pump shall be drained and protected from freezing by heating or other approved means. Where pumps are installed in basements, such pumps shall be mounted on a block or shelf not less than 18 inches (457 mm) above the basement floor. Well pits shall be prohibited.

SECTION 603
WATER SERVICE

603.1 Size of water service pipe. The water service pipe shall be sized to supply water to the structure in the quantities and at the pressures required in this code. The minimum diameter of water service pipe shall be $^3/_4$ inch (19.1 mm).

603.2 Separation of water service and building sewer. Water service pipe and the building sewer shall be separated by 5 feet (1524 mm) of undisturbed or compacted earth.

Exceptions:

1. The required separation distance shall not apply where the bottom of the water service pipe within 5 feet (1524 mm) of the sewer is a minimum of 12 inches (305 mm) above the top of the highest point of the sewer and the pipe materials conform to Section 703.1.

2. Water service pipe is permitted to be located in the same trench with a building sewer, provided such sewer is constructed of materials listed in Table 702.2.

3. The required separation distance shall not apply where a water service pipe crosses a sewer pipe provided the water service pipe is sleeved to at least 5 feet (1524 mm) horizontally from the sewer pipe centerline, on both sides of such crossing with pipe materials listed in Table 605.3, Table 702.2 or Table 702.3.

603.2.1 Water service near sources of pollution. Potable water service pipes shall not be located in, under or above cesspools, septic tanks, septic tank drainage fields or seepage pits (see Section 605.1 for soil and groundwater conditions).

SECTION 604
DESIGN OF BUILDING WATER DISTRIBUTION SYSTEM

604.1 General. The design of the water distribution system shall conform to accepted engineering practice. Methods utilized to determine pipe sizes shall be approved.

604.2 System interconnection. At the points of interconnection between the hot and cold water supply piping systems and the individual fixtures, appliances or devices, provisions shall be made to prevent flow between such piping systems.

604.3 Water distribution system design criteria. The water distribution system shall be designed, and pipe sizes shall be selected such that under conditions of peak demand, the capacities at the fixture supply pipe outlets shall not be less than shown in Table 604.3. The minimum flow rate and flow pressure provided to fixtures and appliances not listed in Table 604.3 shall be in accordance with the manufacturer's installation instructions.

TABLE 604.3
WATER DISTRIBUTION SYSTEM DESIGN CRITERIA
REQUIRED CAPACITY AT FIXTURE SUPPLY PIPE OUTLETS

FIXTURE SUPPLY OUTLET SERVING	FLOW RATE[a] (gpm)	FLOW PRESSURE (psi)
Bathtub	4	8
Bidet	2	4
Combination fixture	4	8
Dishwasher, residential	2.75	8
Drinking fountain	0.75	8
Laundry tray	4	8
Lavatory	2	8
Shower	3	8
Shower, temperature controlled	3	20
Sillcock, hose bibb	5	8
Sink, residential	2.5	8
Sink, service	3	8
Urinal, valve	15	15
Water closet, blow out, flushometer valve	35	25
Water closet, flushometer tank	1.6	15
Water closet, siphonic, flushometer valve	25	15
Water closet, tank, close coupled	3	8
Water closet, tank, one piece	6	20

For SI: 1 pound per square inch = 6.895 kPa, 1 gallon per minute = 3.785 L/m.

a. For additional requirements for flow rates and quantities, see Section 604.4.

604.4 Maximum flow and water consumption. The maximum water consumption flow rates and quantities for all plumbing fixtures and fixture fittings shall be in accordance with Table 604.4.

Exceptions:

1. Blowout design water closets [3.5 gallons (13 L) per flushing cycle].
2. Vegetable sprays.
3. Clinical sinks [4.5 gallons (17 L) per flushing cycle].
4. Service sinks.
5. Emergency showers.

TABLE 604.4
MAXIMUM FLOW RATES AND CONSUMPTION FOR PLUMBING FIXTURES AND FIXTURE FITTINGS

PLUMBING FIXTURE OR FIXTURE FITTING	MAXIMUM FLOW RATE OR QUANTITY[b]
Lavatory, private	2.2 gpm at 60 psi
Lavatory, public, (metering)	0.25 gallon per metering cycle
Lavatory, public (other than metering)	0.5 gpm at 60 psi
Shower head[a]	2.5 gpm at 80 psi
Sink faucet	2.2 gpm at 60 psi
Urinal	1.0 gallon per flushing cycle
Water closet	1.6 gallons per flushing cycle

For SI: 1 gallon = 3.785 L, 1 gallon per minute = 3.785 L/m, 1 pound per square inch = 6.895 kPa.

a. A hand-held shower spray is a shower head.

b. Consumption tolerances shall be determined from referenced standards.

604.5 Size of fixture supply. The minimum size of a fixture supply pipe shall be as shown in Table 604.5. The fixture supply pipe shall not terminate more than 30 inches (762 mm) from the point of connection to the fixture. A reduced-size flexible water connector installed between the supply pipe and the fixture shall be of an approved type. The supply pipe shall extend to the floor or wall adjacent to the fixture. The minimum size of individual distribution lines utilized in parallel water distribution systems shall be as shown in Table 604.5.

604.6 Variable street pressures. Where street water main pressures fluctuate, the building water distribution system shall be designed for the minimum pressure available.

604.7 Inadequate water pressure. Wherever water pressure from the street main or other source of supply is insufficient to provide flow pressures at fixture outlets as required under Table 604.3, a water pressure booster system conforming to Section 606.5 shall be installed on the building water supply system.

604.8 Water-pressure reducing valve or regulator. Where water pressure within a building exceeds 80 psi (552 kPa) static, an approved water-pressure reducing valve conforming to ASSE 1003 with strainer shall be installed to reduce the pressure in the building water distribution piping to 80 psi (552 kPa) static or less.

Exception: Service lines to sill cocks and outside hydrants, and main supply risers where pressure from the mains is reduced to 80 psi (552 kPa) or less at individual fixtures.

604.8.1 Valve design. The pressure-reducing valve shall be designed to remain open to permit uninterrupted water flow in case of valve failure.

TABLE 604.5
MINIMUM SIZES OF FIXTURE WATER SUPPLY PIPES

FIXTURE	MINIMUM PIPE SIZE (inch)
Bathtubs[a] (60″ × 32″ and smaller	$^1/_2$
Bathtubs[a] (larger than 60″ × 32″)	$^1/_2$
Bidet	$^3/_8$
Combination sink and tray	$^1/_2$
Dishwasher, domestic[a]	$^1/_2$
Drinking fountain	$^3/_8$
Hose bibbs	$^1/_2$
Kitchen sink[a]	$^1/_2$
Laundry, 1, 2 or 3 compartments[a]	$^1/_2$
Lavatory	$^3/_8$
Shower, single head[a]	$^1/_2$
Sinks, flushing rim	$^3/_4$
Sinks, service	$^1/_2$
Urinal, flush tank	$^1/_2$
Urinal, flush valve	$^3/_4$
Wall hydrant	$^1/_2$
Water closet, flush tank	$^3/_8$
Water closet, flush valve	1
Water closet, flushometer tank	$^3/_8$
Water closet, one piece[a]	$^1/_2$

For SI: 1 inch = 25.4 mm, 1 foot = 304.8 mm, 1 pound per square inch = 6.895 kPa.

a. Where the developed length of the distribution line is 60 feet or less, and the available pressure at the meter is a minimum of 35 psi, the minimum size of an individual distribution line supplied from a manifold and installed as part of a parallel water distribution system shall be one nominal tube size smaller than the sizes indicated.

604.8.2 Repair and removal. All water-pressure reducing valves, regulators and strainers shall be so constructed and installed as to permit repair or removal of parts without breaking a pipeline or removing the valve and strainer from the pipeline.

604.9 Water hammer. The flow velocity of the water distribution system shall be controlled to reduce the possibility of water hammer. A water-hammer arrestor shall be installed where quick-closing valves are utilized. Water-hammer arrestors shall be installed in accordance with the manufacturer's specifications. Water-hammer arrestors shall conform to ASSE 1010.

604.10 Parallel water distribution system manifolds. Hot water and cold water manifolds installed with parallel connected individual distribution lines to each fixture or fixture fitting shall be designed in accordance with Sections 604.10.1 through 604.10.3.

604.10.1 Manifold sizing. Hot water and cold water manifolds shall be sized in accordance with Table 604.10.1. The total gallons per minute is the demand of all outlets supplied.

TABLE 604.10.1
MANIFOLD SIZING

NOMINAL SIZE INTERNAL DIAMETER (inches)	MAXIMUM DEMAND (gpm)	
	Velocity at 4 feet per second	Velocity at 8 feet per second
$^1/_2$	2	5
$^3/_4$	6	11
1	10	20
$1^1/_4$	15	31
$1^1/_2$	22	44

For SI: 1 inch = 25.4 mm, 1 gallon per minute = 3.785 L/m, 1 foot per second = 0.305 m/s.

604.10.2 Valves. Individual fixture shutoff valves installed at the manifold shall be identified as to the fixture being supplied.

604.10.3 Access. Access shall be provided to manifolds.

604.11 Individual pressure balancing in-line valves for individual fixture fittings. Where individual pressure balancing in-line valves for individual fixture fittings are installed, such valves shall comply with ASSE 1066. Such valves shall be installed in an accessible location and shall not be utilized alone as a substitute for the balanced pressure, thermostatic or combination shower valves required in Section 424.3.

SECTION 605
MATERIALS, JOINTS AND CONNECTIONS

605.1 Soil and ground water. The installation of a water service or water distribution pipe shall be prohibited in soil and ground water contaminated with solvents, fuels, organic compounds or other detrimental materials causing permeation, corrosion, degradation or structural failure of the piping material. Where detrimental conditions are suspected, a chemical analysis of the soil and ground water conditions shall be required to ascertain the acceptability of the water service or water distribution piping material for the specific installation. Where detrimental conditions exist, approved alternative materials or routing shall be required.

605.2 Lead content of water supply pipe and fittings. Pipe and pipe fittings, including valves and faucets, utilized in the water supply system shall have a maximum of 8-percent lead content.

605.3 Water service pipe. Water service pipe shall conform to NSF 61 and shall conform to one of the standards listed in Table 605.3. All water service pipe or tubing, installed underground and outside of the structure, shall have a minimum working pressure rating of 160 psi (1100 kPa) at 73.4°F (23°C). Where the water pressure exceeds 160 psi (1100 kPa), piping material shall have a minimum rated working pressure equal to the highest available pressure. Plastic water service piping shall terminate within 5 feet (1524 mm) inside of the point where the pipe penetrates an exterior wall or slab on grade. All ductile iron water pipe shall be cement mortar lined in accordance with AWWA C104.

605.3.1 Dual check-valve-type backflow preventer. Where a dual check-valve backflow preventer is installed on the water supply system, it shall comply with ASSE 1024.

TABLE 605.3
WATER SERVICE PIPE

MATERIAL	STANDARD
Acrylonitrile butadiene styrene (ABS) plastic pipe	ASTM D 1527; ASTM D 2282
Asbestos-cement pipe	ASTM C 296
Brass pipe	ASTM B 43
Chlorinated polyvinyl chloride (CPVC) plastic pipe	ASTM D 2846; ASTM F 441; ASTM F 442; CSA B137.6
Copper or copper-alloy pipe	ASTM B 42; ASTM B 302
Copper or copper-alloy tubing (Type K, WK, L, WL, M or WM)	ASTM B 75; ASTM B 88; ASTM B 251; ASTM B 447
Cross-linked polyethylene (PEX) plastic tubing	ASTM F 876; ASTM F 877; CSA-B137.5
Cross-linked polyethylene/aluminum/cross-linked polyethylene (PEX-AL-PEX) pipe	ASTM F 1281; CAN/CSA B137.10M
Ductile iron water pipe	AWWA C151; AWWA C115
Galvanized steel pipe	ASTM A 53
Polybutylene (PB) plastic pipe and tubing	ASTM D 2662; ASTM D 2666; ASTM D 3309; CAN3-B137.8M
Polyethylene (PE) plastic pipe	ASTM D 2239; CSA-B137.1
Polyethylene (PE) plastic tubing	ASTM D 2737; CSA B137.1
Polyethylene/aluminum/polethylene (PE-AL-PE) pipe	ASTM F 1282 CAN/CSA-B137.9
Polyvinyl chloride (PVC) plastic pipe	ASTM D 1785; ASTM D 2241; ASTM D 2672; CSA-B137.3
Stainless steel pipe (Type 304/304L)	ASTM A 312; ASTM A 778
Stainless steel pipe (Type 316/316L)	ASTM A 312; ASTM A 778

TABLE 605.4
WATER DISTRIBUTION PIPE

MATERIAL	STANDARD
Brass pipe	ASTM B 43
Chlorinated polyvinyl chloride (CPVC) plastic pipe and tubing	ASTM D 2846; ASTM F 441; ASTM F 442; CSA B137.6
Copper or copper-alloy pipe	ASTM B 42; ASTM B 302
Copper or copper-alloy tubing (Type K, WK, L, WL, M or WM)	ASTM B 75; ASTM B 88; ASTM B 251; ASTM B 447
Cross-linked polyethylene (PEX) plastic tubing	ASTM F 877; CSA-B137.5
Cross-linked polyethylene/aluminum/cross-linked polyethylene (PEX-AL-PEX) pipe	ASTM F 1281; CAN/CSA-B137.10M
Galvanized steel pipe	ASTM A 53
Polybutylene (PB) plastic pipe and tubing	ASTM D 3309; CAN3-B137.8M
Polyethylene/Aluminum/Polyethylene (PE-AL-PE) composite pipe	ASTM F 1282
Stainless steel pipe (Type 304/304L)	ASTM A 312; ASTM A 778
Stainless steel pipe (Type 316/316L)	ASTM A 312; ASTM A 778

TABLE 605.5
PIPE FITTINGS

MATERIAL	STANDARD
Acrylonitrile butadiene styrene (ABS) plastic	ASTM D 2468
Cast-iron	ASME B16.4; ASME B16.12
Chlorinated polyvinyl chloride (CPVC) plastic	ASTM F 437; ASTM F 438; ASTM F 439
Copper or copper alloy	ASME B16.15; ASME B16.18; ASME B16.22; ASME B16.23; ASME B16.26; ASME B16.29
Fittings for cross-linked polyethylene (PEX) plastic tubing	ASTM F 1807, ASTM F 1960, ASTM F 2080
Gray iron and ductile iron	AWWA C 110; AWWA C 153
Malleable iron	ASME B16.3
Metal (brass) insert fittings for Polyethylene/Aluminum/Polyethylene (PE-AL-PE) and Cross-linked Polyethylene/Aluminum/Polyethylene (PEX-AL-PEX)	ASTM F 1974
Polyethylene (PE) plastic	ASTM D 2609
Polyvinyl chloride (PVC) plastic	ASTM D 2464; ASTM D 2466; ASTM D 2467; CSA-B137.2
Stainless steel (Type 304/304L)	ASTM A 312; ASTM A 778
Stainless steel (Type 316/316L)	ASTM A 312; ASTM A 778
Steel	ASME B16.9; ASME B16.11; ASME B16.28

605.4 Water distribution pipe. Water distribution pipe shall conform to NSF 61 and shall conform to one of the standards listed in Table 605.4. All hot water distribution pipe and tubing shall have a minimum pressure rating of 100 psi (690 kPa) at 180°F (82°C).

605.5 Fittings. Pipe fittings shall be approved for installation with the piping material installed and shall conform to the respective pipe standards or one of the standards listed in Table 605.5. All pipe fittings utilized in water supply systems shall also conform to NSF 61. The fittings shall not have ledges, shoulders or reductions capable of retarding or obstructing flow in the piping. Ductile and gray iron pipe fittings shall be cement mortar lined in accordance with AWWA C104.

605.5.1 Mechanically formed tee fittings. Mechanically extracted outlets shall have a height not less than three times the thickness of the branch tube wall.

605.5.1.1 Full flow assurance. Branch tubes shall not restrict the flow in the run tube. A dimple/depth stop shall be formed in the branch tube to ensure that penetration into the collar is of the correct depth. For inspection purposes, a second dimple shall be placed 0.25 inch (6.4 mm) above the first dimple. Dimples shall be aligned with the tube run.

605.5.1.2 Brazed joints. Mechanically formed tee fittings shall be brazed in accordance with Section 605.14.1.

605.6 Flexible water connectors. Flexible water connectors exposed to continuous pressure shall conform to ASME A112.18.6. Access shall be provided to all flexible water connectors.

605.7 Valves. All valves shall be of the approved type and compatible with the type of piping material installed in the system.

605.8 Manufactured pipe nipples. Manufactured pipe nipples shall conform to one of the standards listed in Table 605.8.

TABLE 605.8
MANUFACTURED PIPE NIPPLES

MATERIAL	STANDARD
Brass-, copper-, chromium-plated	ASTM B 687
Steel	ASTM A 733

605.9 Prohibited joints and connections. The following types of joints and connections shall be prohibited:

1. Cement or concrete joints.
2. Joints made with fittings not approved for the specific installation.
3. Solvent-cement joints between different types of plastic pipe.
4. Saddle-type fittings.

605.10 ABS plastic. Joints between ABS plastic pipe or fittings shall comply with Sections 605.10.1 through 605.10.3.

605.10.1 Mechanical joints. Mechanical joints on water pipes shall be made with an elastomeric seal conforming to ASTM D 3139. Mechanical joints shall only be installed in underground systems, unless otherwise approved. Joints shall be installed only in accordance with the manufacturer's instructions.

605.10.2 Solvent cementing. Joint surfaces shall be clean and free from moisture. Solvent cement that conforms to ASTM D 2235 shall be applied to all joint surfaces. The joint shall be made while the cement is wet. Joints shall be made in accordance with ASTM D 2235. Solvent-cement joints shall be permitted above or below ground.

605.10.3 Threaded joints. Threads shall conform to ASME B1.20.1. Schedule 80 or heavier pipe shall be permitted to be threaded with dies specifically designed for plastic pipe. Approved thread lubricant or tape shall be applied on the male threads only.

605.11 Asbestos-cement. Joints between asbestos-cement pipe or fittings shall be made with a sleeve coupling of the same composition as the pipe, sealed with an elastomeric ring conforming to ASTM D 1869.

605.12 Brass. Joints between brass pipe or fittings shall comply with Sections 605.12.1 through 605.12.4.

605.12.1 Brazed joints. All joint surfaces shall be cleaned. An approved flux shall be applied where required. The joint shall be brazed with a filler metal conforming to AWS A5.8.

605.12.2 Mechanical joints. Mechanical joints shall be installed in accordance with the manufacturer's instructions.

605.12.3 Threaded joints. Threads shall conform to ASME B1.20.1. Pipe-joint compound or tape shall be applied on the male threads only.

605.12.4 Welded joints. All joint surfaces shall be cleaned. The joint shall be welded with an approved filler metal.

605.13 Gray iron and ductile iron joints. Joints for gray and ductile iron pipe and fittings shall comply with AWWA C111 and shall be installed in accordance with the manufacturer's installation instructions.

605.14 Copper pipe. Joints between copper or copper-alloy pipe or fittings shall comply with Sections 605.14.1 through 605.14.5.

605.14.1 Brazed joints. All joint surfaces shall be cleaned. An approved flux shall be applied where required. The joint shall be brazed with a filler metal conforming to AWS A5.8.

605.14.2 Mechanical joints. Mechanical joints shall be installed in accordance with the manufacturer's instructions.

605.14.3 Soldered joints. Solder joints shall be made in accordance with the methods of ASTM B 828. All cut tube ends shall be reamed to the full inside diameter of the tube end. All joint surfaces shall be cleaned. A flux conforming to ASTM B 813 shall be applied. The joint shall be soldered with a solder conforming to ASTM B 32. The joining of water supply piping shall be made with lead-free solder and fluxes. "Lead free" shall mean a chemical composition equal to or less than 0.2-percent lead.

605.14.4 Threaded joints. Threads shall conform to ASME B1.20.1. Pipe-joint compound or tape shall be applied on the male threads only.

605.14.5 Welded joints. All joint surfaces shall be cleaned. The joint shall be welded with an approved filler metal.

605.15 Copper tubing. Joints between copper or copper-alloy tubing or fittings shall comply with Sections 605.15.1 through 605.15.4.

605.15.1 Brazed joints. All joint surfaces shall be cleaned. An approved flux shall be applied where required. The joint

shall be brazed with a filler metal conforming to AWS A5.8.

605.15.2 Flared joints. Flared joints for water pipe shall be made by a tool designed for that operation.

605.15.3 Mechanical joints. Mechanical joints shall be installed in accordance with the manufacturer's instructions.

605.15.4 Soldered joints. Solder joints shall be made in accordance with the methods of ASTM B 828. All cut tube ends shall be reamed to the full inside diameter of the tube end. All joint surfaces shall be cleaned. A flux conforming to ASTM B 813 shall be applied. The joint shall be soldered with a solder conforming to ASTM B 32. The joining of water supply piping shall be made with lead-free solders and fluxes. "Lead free" shall mean a chemical composition equal to or less than 0.2-percent lead.

605.16 CPVC plastic. Joints between CPVC plastic pipe or fittings shall comply with Sections 605.16.1 through 605.16.3.

605.16.1 Mechanical joints. Mechanical joints shall be installed in accordance with the manufacturer's instructions.

605.16.2 Solvent cementing. Joint surfaces shall be clean and free from moisture, and an approved primer shall be applied. Solvent cement, orange in color and conforming to ASTM F 493, shall be applied to all joint surfaces. The joint shall be made while the cement is wet, and in accordance with ASTM D 2846 or ASTM F 493. Solvent-cement joints shall be permitted above or below ground.

Exception: A primer is not required where all of the following conditions apply:

1. The solvent cement used is third-party certified as conforming to ASTM F 493.

2. The solvent cement used is yellow in color.

3. The solvent cement is used only for joining ½ inch (12.7 mm) through 2 inch (51 mm) diameter CPVC pipe and fittings.

4. The CPVC pipe and fittings are manufactured in accordance with ASTM D 2846.

605.16.3 Threaded joints. Threads shall conform to ASME B1.20.1. Schedule 80 or heavier pipe shall be permitted to be threaded with dies specifically designed for plastic pipe, but the pressure rating of the pipe shall be reduced by 50 percent. Thread by socket molded fittings shall be permitted. Approved thread lubricant or tape shall be applied on the male threads only.

605.17 Cross-linked polyethylene plastic. Joints between cross-linked polyethylene plastic tubing or fittings shall comply with Sections 605.17.1 and 605.17.2.

605.17.1 Flared joints. Flared pipe ends shall be made by a tool designed for that operation.

605.17.2 Mechanical joints. Mechanical joints shall be installed in accordance with the manufacturer's instructions. Fittings for cross-linked polyethylene (PEX) plastic tubing as described in ASTM F 1807, ASTM F 1960 and ASTM F 2080 shall be installed in accordance with the manufacturer's instructions.

605.18 Steel. Joints between galvanized steel pipe or fittings shall comply with Sections 605.18.1 and 605.18.2.

605.18.1 Threaded joints. Threads shall conform to ASME B1.20.1. Pipe-joint compound or tape shall be applied on the male threads only.

605.18.2 Mechanical joints. Joints shall be made with an approved elastomeric seal. Mechanical joints shall be installed in accordance with the manufacturer's instructions.

605.19 Polybutylene plastic. Joints between polybutylene plastic pipe and tubing or fittings shall comply with Sections 605.19.1 through 605.19.3.

605.19.1 Flared joints. Flared pipe ends shall be made by a tool designed for that operation.

605.19.2 Heat-fusion joints. Joints shall be of the socket-fusion or butt-fusion type. Joint surfaces shall be clean and free from moisture. All joint surfaces shall be heated to melt temperature and joined. The joint shall be undisturbed until cool. Joints shall be made in accordance with ASTM D 2657, ASTM D 3309 or CAN3-B137.8M.

605.19.3 Mechanical joints. Mechanical joints shall be installed in accordance with the manufacturer's instructions. Metallic lock rings employed with insert fittings as described in ASTM D 3309 or CAN3-B137.8M shall be installed in accordance with the manufacturer's instructions.

605.20 Polyethylene plastic. Joints between polyethylene plastic pipe and tubing or fittings shall comply with Sections 605.20.1 through 605.20.4.

605.20.1 Flared joints. Flared joints shall be permitted where so indicated by the pipe manufacturer. Flared joints shall be made by a tool designed for that operation.

605.20.2 Heat-fusion joints. Joint surfaces shall be clean and free from moisture. All joint surfaces shall be heated to melt temperature and joined. The joint shall be undisturbed until cool. Joints shall be made in accordance with ASTM D 2657.

605.20.3 Mechanical joints. Mechanical joints shall be installed in accordance with the manufacturer's instructions.

605.20.4 Installation. Polyethylene pipe shall be cut square, with a cutter designed for plastic pipe. Except where joined by heat fusion, pipe ends shall be chamfered to remove sharp edges. Kinked pipe shall not be installed. The minimum pipe bending radius shall not be less than 30 pipe diameters, or the minimum coil radius, whichever is greater. Piping shall not be bent beyond straightening of the curvature of the coil. Bends shall not be permitted within 10 pipe diameters of any fitting or valve. Stiffener inserts installed with compression-type couplings and fittings shall not extend beyond the clamp or nut of the coupling or fitting.

605.21 PVC plastic. Joints between PVC plastic pipe or fittings shall comply with Sections 605.21.1 through 605.21.3.

605.21.1 Mechanical joints. Mechanical joints on water pipe shall be made with an elastomeric seal conforming to ASTM D 3139. Mechanical joints shall not be installed in above-ground systems unless otherwise approved. Joints shall be installed in accordance with the manufacturer's instructions.

605.21.2 Solvent cementing. Joint surfaces shall be clean and free from moisture. A purple primer that conforms to ASTM F 656 shall be applied. Solvent cement not purple in color and conforming to ASTM D 2564 or CSA-B137.3 shall be applied to all joint surfaces. The joint shall be made while the cement is wet and shall be in accordance with ASTM D 2855. Solvent-cement joints shall be permitted above or below ground.

605.21.3 Threaded joints. Threads shall conform to ASME B1.20.1. Schedule 80 or heavier pipe shall be permitted to be threaded with dies specifically designed for plastic pipe, but the pressure rating of the pipe shall be reduced by 50 percent. Thread by socket molded fittings shall be permitted. Approved thread lubricant or tape shall be applied on the male threads only.

605.22 Stainless steel. Joints between stainless steel pipe and fittings shall comply with Sections 605.22.1 and 605.22.2.

605.22.1 Mechanical joints. Mechanical joints shall be installed in accordance with the manufacturer's instructions.

605.22.2 Welded joints. All joint surfaces shall be cleaned. The joint shall be welded autogenously or with an approved filler metal as referenced in ASTM A 312.

605.23 Joints between different materials. Joints between different piping materials shall be made with a mechanical joint of the compression or mechanical-sealing type, or as permitted in Sections 605.23.1, 605.23.2 and 605.23.3. Connectors or adapters shall have an elastomeric seal conforming to ASTM D 1869 or ASTM F 477. Joints shall be installed in accordance with the manufacturer's instructions.

605.23.1 Copper or copper-alloy tubing to galvanized steel pipe. Joints between copper or copper-alloy tubing and galvanized steel pipe shall be made with a brass fitting or dielectric fitting. The copper tubing shall be soldered to the fitting in an approved manner, and the fitting shall be screwed to the threaded pipe.

605.23.2 Plastic pipe or tubing to other piping material. Joints between different grades of plastic pipe or between plastic pipe and other piping material shall be made with an approved adapter fitting.

605.23.3 Stainless steel. Joints between stainless steel and different piping materials shall be made with a mechanical joint of the compression or mechanical sealing type or a dielectric fitting.

SECTION 606
INSTALLATION OF THE BUILDING
WATER DISTRIBUTION SYSTEM

606.1 Location of full-open valves. Full-open valves shall be installed in the following locations:

1. On the building water service pipe from the public water supply near the curb.

2. On the water distribution supply pipe at the entrance into the structure.

3. On the discharge side of every water meter.

4. On the base of every water riser pipe in occupancies other than multiple-family residential occupancies that are two stories or less in height and in one- and two-family residential occupancies.

5. On the top of every water down-feed pipe in occupancies other than one- and two-family residential occupancies.

6. On the entrance to every water supply pipe to a dwelling unit, except where supplying a single fixture equipped with individual stops.

7. On the water supply pipe to a gravity or pressurized water tank.

8. On the water supply pipe to every water heater.

606.2 Location of shutoff valves. Shutoff valves shall be installed in the following locations:

1. On the fixture supply to each plumbing fixture other than bathtubs and showers in one- and two-family residential occupancies, and other than in individual guestrooms that are provided with unit shutoff valves in hotels, motels, boarding houses and similar occupancies.

2. On the water supply pipe to each sillcock.

3. On the water supply pipe to each appliance or mechanical equipment.

606.3 Access to valves. Access shall be provided to all required full-open valves and shutoff valves.

606.4 Valve identification. Service and hose bibb valves shall be identified. All other valves installed in locations that are not adjacent to the fixture or appliance shall be identified, indicating the fixture or appliance served.

606.5 Water pressure booster systems. Water pressure booster systems shall be provided as required by Sections 606.5.1 through 606.5.10.

606.5.1 Water pressure booster systems required. Where the water pressure in the public water main or individual water supply system is insufficient to supply the minimum pressures and quantities specified in this code, the supply shall be supplemented by an elevated water tank, a hydropneumatic pressure booster system or a water pressure booster pump installed in accordance with Section 606.5.5.

606.5.2 Support. All water supply tanks shall be supported in accordance with the *International Building Code.*

606.5.3 Covers. All water supply tanks shall be covered to keep out unauthorized persons, dirt and vermin. The covers of gravity tanks shall be vented with a return bend vent pipe with an area not less than the area of the down-feed riser pipe, and the vent shall be screened with a corrosion-resistant screen of not less than 16 by 20 mesh per inch (630 by 787 mesh per m).

606.5.4 Overflows for water supply tanks. Each gravity or suction water supply tank shall be provided with an overflow with a diameter not less than that shown in Table 606.5.4. The overflow outlet shall discharge above and within not less than 6 inches (152 mm) of a roof or roof drain, floor or floor drain, or over an open water-supplied fixture. The overflow outlet shall be covered with a corrosion-resistant screen of not less than 16 by 20 mesh per inch (630 by 787 mesh per m) and by 0.25-inch (6.4 mm) hardware cloth or shall terminate in a horizontal angle seat check valve. Drainage from overflow pipes shall be directed so as not to freeze on roof walks.

TABLE 606.5.4
SIZES FOR OVERFLOW PIPES FOR WATER SUPPLY TANKS

MAXIMUM CAPACITY OF WATER SUPPLY LINE TO TANK (gpm)	DIAMETER OF OVERFLOW PIPE (inches)
0 - 50	2
50 - 150	$2^1/_2$
150 - 200	3
200 - 400	4
400 - 700	5
700 - 1,000	6
Over 1,000	8

For SI: 1 inch = 25.4 mm, 1 gallon per minute = 3.785 L/m.

606.5.5 Low-pressure cutoff required on booster pumps. A low-pressure cutoff shall be installed on all booster pumps in a water pressure booster system to prevent creation of a vacuum or negative pressure on the suction side of the pump when a positive pressure of 10 psi (68.94 kPa) or less occurs on the suction side of the pump.

606.5.6 Potable water inlet control and location. Potable water inlets to gravity tanks shall be controlled by a fill valve or other automatic supply valve installed so as to prevent the tank from overflowing. The inlet shall be terminated so as to provide an air gap not less than 4 inches (102 mm) above the overflow.

606.5.7 Tank drain pipes. A valved pipe shall be provided at the lowest point of each tank to permit emptying of the tank. The tank drain pipe shall discharge as required for overflow pipes and shall not be smaller in size than specified in Table 606.5.7.

606.5.8 Prohibited location of potable supply tanks. Potable water gravity tanks or manholes of potable water pressure tanks shall not be located directly under any soil or waste piping or any source of contamination.

606.5.9 Pressure tanks, vacuum relief. All water pressure tanks shall be provided with a vacuum relief valve at the top of the tank that will operate up to a maximum water pressure of 200 psi (1380 kPa) and up to a maximum temperature of 200°F (93°C). The minimum size of such vacuum relief valve shall be 0.50 inch (12.7 mm).

Exception: This section shall not apply to pressurized captive air diaphragm/bladder tanks.

TABLE 606.5.7
SIZE OF DRAIN PIPES FOR WATER TANKS

TANK CAPACITY (gallons)	DRAIN PIPE (inches)
Up to 750	1
751 to 1,500	$1^1/_2$
1,501 to 3,000	2
3,001 to 5,000	$2^1/_2$
5,000 to 7,500	3
Over 7,500	4

For SI: 1 inch = 25.4 mm, 1 gallon = 3.785 L.

606.5.10 Pressure relief for tanks. Every pressure tank in a hydropneumatic pressure booster system shall be protected with a pressure relief valve. The pressure relief valve shall be set at a maximum pressure equal to the rating of the tank. The relief valve shall be installed on the supply pipe to the tank or on the tank. The relief valve shall discharge by gravity to a safe place of disposal.

606.6 Water supply system test. Upon completion of a section of or the entire water supply system, the system, or portion completed, shall be tested in accordance with Section 312.

SECTION 607
HOT WATER SUPPLY SYSTEM

607.1 Where required. In residential occupancies, hot water shall be supplied to all plumbing fixtures and equipment utilized for bathing, washing, culinary purposes, cleansing, laundry or building maintenance. In nonresidential occupancies, hot water shall be supplied to all plumbing fixtures and equipment utilized for culinary purposes, cleansing, laundry or building maintenance. In nonresidential occupancies, hot water or tempered water shall be supplied for bathing and washing purposes. Tempered water shall be delivered from accessible hand-washing facilities.

607.2 Hot water supply temperature maintenance. Where the developed length of hot water piping from the source of hot water supply to the farthest fixture exceeds 100 feet (30 480 mm), the hot water supply system shall be provided with a method of maintaining the temperature in accordance with the *International Energy Conservation Code.*

607.2.1 Piping insulation. Circulating hot water system piping shall be insulated in accordance with the *International Energy Conservation Code.*

[E] 607.2.2 Hot water system controls. Automatic circulating hot water system pumps or heat trace shall be arranged to be conveniently turned off, automatically or manually, when the hot water system is not in operation.

607.2.3 Recirculating pump. Where a thermostatic mixing valve is used in a system with a hot water recirculating pump, the hot water or tempered water return line shall be routed to the cold water inlet pipe of the water heater and the cold water inlet pipe or the hot water return connection of the thermostatic mixing valve.

607.3 Thermal expansion control. A means of controlling increased pressure caused by thermal expansion shall be provided where required in accordance with Sections 607.3.1 and 607.3.2.

607.3.1 Pressure-reducing valve. For water service system sizes up to and including 2 inches (51 mm), a device for controlling pressure shall be installed where, because of thermal expansion, the pressure on the downstream side of a pressure-reducing valve exceeds the pressure-reducing valve setting.

607.3.2 Backflow prevention device or check valve. Where a backflow prevention device, check valve or other device is installed on a water supply system utilizing storage water heating equipment such that thermal expansion causes an increase in pressure, a device for controlling pressure shall be installed.

607.4 Flow of hot water to fixtures. Fixture fittings, faucets and diverters shall be installed and adjusted so that the flow of hot water from the fittings corresponds to the left-hand side of the fixture fitting.

Exception: Shower and tub/shower mixing valves conforming to ASSE 1016, where the flow of hot water corresponds to the markings on the device.

SECTION 608
PROTECTION OF POTABLE WATER SUPPLY

608.1 General. A potable water supply system shall be designed, installed and maintained in such a manner so as to prevent contamination from nonpotable liquids, solids or gases being introduced into the potable water supply through cross-connections or any other piping connections to the system. Backflow preventer applications shall conform to Table 608.1, except as specifically stated in Sections 608.2 through 608.16.9.

608.2 Plumbing fixtures. The supply lines or fittings for every plumbing fixture shall be installed so as to prevent backflow.

608.3 Devices, appurtenances, appliances and apparatus. All devices, appurtenances, appliances and apparatus intended to serve some special function, such as sterilization, distillation, processing, cooling, or storage of ice or foods, and that connect to the water supply system, shall be provided with protection against backflow and contamination of the water supply system. Water pumps, filters, softeners, tanks and all other appliances and devices that handle or treat potable water shall be protected against contamination.

608.3.1 Special equipment, water supply protection. The water supply for hospital fixtures shall be protected against backflow with a reduced pressure principle backflow preventer, an atmospheric or spill-proof vacuum breaker, or an air gap. Vacuum breakers for bedpan washer hoses shall not be located less than 5 feet (1524 mm) above the floor. Vacuum breakers for hose connections in health care or laboratory areas shall not be less than 6 feet (1829 mm) above the floor.

608.4 Water service piping. Water service piping shall be protected in accordance with Sections 603.2 and 603.2.1.

TABLE 608.1
APPLICATION OF BACKFLOW PREVENTERS

DEVICE	DEGREE OF HAZARD [a]	APPLICATION [b]	APPLICABLE STANDARDS
Air gap	High or low hazard	Backsiphonage or backpressure	ASME A112.1.2
Air gap fittings for use with plumbing fixtures, appliances and appurtenances	High or low hazard	Backsiphonage or backpressure	ASME A112.1.3
Antisiphon-type fill valves for gravity water closet flush tanks	High hazard ·	Backsiphonage only	ASSE 1002, CSA-B125
Barometric loop	High or low hazard	Backsiphonage only	(See Section 608.13.4)
Reduced pressure principle backflow preventer and reduced pressure principle fire protection backflow preventer	High or low hazard	Backpressure or backsiphonage Sizes $^3/_8$"- 16"	ASSE 1013, AWWA C511, CAN/CSA B64.4
Reduced pressure detector fire protection backflow prevention assemblies	High or low hazard	Backsiphonage or backpressure (Fire sprinkler systems)	ASSE 1047
Double check backflow prevention assembly and double check fire protection backflow prevention assembly	Low hazard	Backpressure or backsiphonage Sizes $^3/_8$" - 16"	ASSE 1015, AWWA C510
Double check detector fire protection backflow prevention assemblies	Low hazard	Backpressure or backsiphonage (Fire sprinkler systems) Sizes 2" - 16"	ASSE 1048
Dual-check-valve-type backflow preventer	Low hazard	Backpressure or backsiphonage Sizes $^1/_4$" - 1"	ASSE 1024
Backflow preventer with intermediate atmospheric vents	Low hazard	Backpressure or backsiphonage Sizes $^1/_4$" - $^3/_4$"	ASSE 1012, CAN/CSA-B64.3
Backflow preventer for carbonated beverage machines	Low hazard	Backpressure or backsiphonage Sizes $^1/_4$"- $^3/_8$"	ASSE 1022
Pipe-applied atmospheric-type vacuum breaker	High or low hazard	Backsiphonage only Sizes $^1/_4$" - 4"	ASSE 1001, CAN/CSA-B64.1.1
Pressure vacuum breaker assembly	High or low hazard	Backsiphonage only Sizes $^1/_2$" - 2"	ASSE 1020
Hose-connection vacuum breaker	High or low hazard	Low head backpressure or backsiphonage Sizes $^1/_2$", $^3/_4$", 1"	ASSE 1011, CAN/CSA-B64.2
Vacuum breaker wall hydrants, frost-resistant, automatic draining type	High or low hazard	Low head backpressure or backsiphonage Sizes $^3/_4$", 1"	ASSE 1019, CAN/CSA-B64.2.2
Laboratory faucet backflow preventer	High or low hazard	Low head backpressure and backsiphonage	ASSE 1035, CSA B64.7
Hose connection backflow preventer	High or low hazard	Low head backpressure, rated working pressure backpressure or backsiphonage Sizes $^1/_2$"-1"	ASSE 1052
Spillproof vacuum breaker	High or low hazard	Backsiphonage only Sizes $^1/_4$"-2"	ASSE 1056

For SI: 1 inch = 25.4 mm.

a. Low hazard–See Pollution (Section 202).
   High hazard–See Contamination (Section 202).

b. See Backpressure (Section 202).
   See Backpressure, low head (Section 202).
   See Backsiphonage (Section 202).

608.5 Chemicals and other substances. Chemicals and other substances that produce either toxic conditions, taste, odor or discoloration in a potable water system shall not be introduced into, or utilized in, such systems.

608.6 Cross-connection control. Cross connections shall be prohibited, except where approved protective devices are installed.

608.6.1 Private water supplies. Cross connections between a private water supply and a potable public supply shall be prohibited.

608.7 Stop-and-waste valves prohibited. Combination stop-and-waste valves or cocks shall not be installed underground.

608.8 Identification of potable and nonpotable water. In all buildings where two or more water distribution systems, one potable water and the other nonpotable water, are installed, each system shall be identified either by color marking or metal tags in accordance with Sections 608.8.1 through 608.8.3.

608.8.1 Information. Pipe identification shall include the contents of the piping system and an arrow indicating the direction of flow. Hazardous piping systems shall also contain information addressing the nature of the hazard. Pipe identification shall be repeated at maximum intervals of 25 feet (7620 mm) and at each point where the piping passes through a wall, floor or roof. Lettering shall be readily observable within the room or space the piping is located.

608.8.2 Color. The color of the pipe identification shall be discernable and consistent throughout the building.

608.8.3 Size. The size of the background color field and lettering shall comply with Table 608.8.3.

TABLE 608.8.3
SIZE OF PIPE IDENTIFICATION

PIPE DIAMETER (inches)	LENGTH BACKGROUND COLOR FIELD (inches)	SIZE OF LETTERS (inches)
¾ to 1¼	8	0.5
1½ to 2	8	0.75
2½ to 6	12	1.25
8 to 10	24	2.5
over 10	32	3.5

For SI: 1 inch = 25.4 mm.

608.9 Reutilization prohibited. Water utilized for the cooling of equipment or other processes shall not be returned to the potable water system. Such water shall be discharged into a drainage system through an air gap or shall be utilized for nonpotable purposes.

608.10 Reuse of piping. Piping that has been utilized for any purpose other than conveying potable water shall not be utilized for conveying potable water.

608.11 Painting of water tanks. The interior surface of a potable water tank shall not be lined, painted or repaired with any material that changes the taste, odor, color or potability of the water supply when the tank is placed in, or returned to, service.

608.12 Pumps and other appliances. Water pumps, filters, softeners, tanks and all other devices that handle or treat potable water shall be protected against contamination.

608.13 Backflow protection. Means of protection against backflow shall be provided in accordance with Sections 608.13.1 through 608.13.9.

608.13.1 Air gap. The minimum required air gap shall be measured vertically from the lowest end of a potable water outlet to the flood level rim of the fixture or receptacle into which such potable water outlet discharges. Air gaps shall comply with ASME A112.1.2 and air gap fittings shall comply with ASME A112.1.3.

608.13.2 Reduced pressure principle backflow preventers. Reduced pressure principle backflow preventers shall conform to ASSE 1013, AWWA C511 or CAN/CSA-B64.3. Reduced pressure detector assembly backflow preventers shall conform to ASSE 1047. These devices shall be permitted to be installed where subject to continuous pressure conditions. The relief opening shall discharge by air gap and shall be prevented from being submerged.

608.13.3 Backflow preventer with intermediate atmospheric vent. Backflow preventers with intermediate atmospheric vents shall conform to ASSE 1012 or CAN/CSA-B64.3. These devices shall be permitted to be installed where subject to continuous pressure conditions. The relief opening shall discharge by air gap and shall be prevented from being submerged.

608.13.4 Barometric loop. Barometric loops shall precede the point of connection and shall extend vertically to a height of 35 feet (10 668 mm). A barometric loop shall only be utilized as an atmospheric-type or pressure-type vacuum breaker.

608.13.5 Pressure-type vacuum breakers. Pressure-type vacuum breakers shall conform to ASSE 1020 and spillproof vacuum breakers shall comply with ASSE 1056. These devices are designed for installation under continuous pressure conditions when the critical level is installed at the required height. Pressure-type vacuum breakers shall not be installed in locations where spillage could cause damage to the structure.

608.13.6 Atmospheric-type vacuum breakers. Pipe-applied atmospheric-type vacuum breakers shall conform to ASSE 1001 or CAN/CSA-B64.1.1. Hose-connection vacuum breakers shall conform to ASSE 1011, ASSE 1019, ASSE 1035, ASSE 1052, CAN/CSA-B64.2, CAN/CSA-B64.2.2 or CSA B64.7. These devices shall operate under normal atmospheric pressure when the critical level is installed at the required height.

608.13.7 Double check-valve assemblies. Double check-valve assemblies shall conform to ASSE 1015 or AWWA C510. Double-detector check-valve assemblies shall conform to ASSE 1048. These devices shall be capable of operating under continuous pressure conditions.

608.13.8 Spillproof vacuum breakers. Spillproof vacuum breakers (SVB) shall conform to ASSE 1056. These devices are designed for installation under continuous-pressure conditions when the critical level is installed at the required height.

608.13.9 Chemical dispenser backflow devices. Backflow devices for chemical dispensers shall comply with ASSE 1055 or shall be equipped with an air gap fitting.

608.14 Location of backflow preventers. Access shall be provided to backflow preventers as specified by the installation instructions of the approved manufacturer.

608.14.1 Outdoor enclosures for backflow prevention devices. Outdoor enclosures for backflow prevention devices shall comply with ASSE 1060.

608.15 Protection of potable water outlets. All potable water openings and outlets shall be protected against backflow in accordance with Section 608.15.1, 608.15.2, 608.15.3, 608.15.4, 608.15.4.1, 608.15.4.2 or 608.15.4.3.

608.15.1 Protection by air gap. Openings and outlets shall be protected by an air gap between the opening and the fixture flood level rim as specified in Table 608.15.1. Openings and outlets equipped for hose connection shall be protected by means other than an air gap.

608.15.2 Protection by a reduced pressure principle backflow preventer. Openings and outlets shall be protected by a reduced pressure principle backflow preventer.

608.15.3 Protection by a backflow preventer with intermediate atmospheric vent. Openings and outlets shall be protected by a backflow preventer with an intermediate atmospheric vent.

608.15.4 Protection by a vacuum breaker. Openings and outlets shall be protected by atmospheric-type or pressure-type vacuum breakers. The critical level of the vacuum breaker shall be set a minimum of 6 inches (152 mm) above the flood level rim of the fixture or device. Fill valves shall be set in accordance with Section 425.3.1. Vacuum breakers shall not be installed under exhaust hoods or similar locations that will contain toxic fumes or vapors. Pipe-applied vacuum breakers shall be installed not less than 6 inches (152 mm) above the flood level rim of the fixture, receptor or device served.

608.15.4.1 Deck-mounted and integral vacuum breakers. Approved deck-mounted or equipment-mounted vacuum breakers and faucets with integral atmospheric or spillproof vacuum breakers shall be installed in accordance with the manufacturer's instructions and the requirements for labeling with the critical level not less than 1 inch (25 mm) above the flood level rim.

608.15.4.2 Hose connections. Sillcocks, hose bibbs, wall hydrants and other openings with a hose connection shall be protected by an atmospheric-type or pressure-type vacuum breaker or a permanently attached hose connection vacuum breaker.

Exceptions:

1. This section shall not apply to water heater and boiler drain valves that are provided with hose connection threads and that are intended only for tank or vessel draining.

2. This section shall not apply to water supply valves intended for connection of clothes washing machines where backflow prevention is otherwise provided or is integral with the machine.

608.16 Connections to the potable water system. Connections to the potable water system shall conform to Sections 608.16.1 through 608.16.9.

TABLE 608.15.1
MINIMUM REQUIRED AIR GAPS

FIXTURE	MINIMUM AIR GAP	
	Away from a wall[a] (inches)	Close to a wall (inches)
Lavatories and other fixtures with effective opening not greater than ½ inch in diameter	1	1½
Sink, laundry trays, gooseneck back faucets and other fixtures with effective openings not greater than ¾ inch in diameter	1.5	2.5
Over-rim bath fillers and other fixtures with effective openings not greater than 1 inch in diameter	2	3
Drinking water fountains, single orifice not greater than $\frac{7}{16}$ inch in diameter or multiple orifices with a total area of 0.150 square inch (area of circle $\frac{7}{16}$ inch in diameter)	1	1½
Effective openings greater than 1 inch	Two times the diameter of the effective opening	Three times the diameter of the effective opening

For SI: 1 inch = 25.4 mm.

a. Applicable where walls or obstructions are spaced from the nearest inside-edge of the spout opening a distance greater than three times the diameter of the effective opening for a single wall, or a distance greater than four times the diameter of the effective opening for two intersecting walls.

608.16.1 Beverage dispensers. The water supply connection to carbonated beverage dispensers shall be protected against backflow by a backflow preventer conforming to ASSE 1022 or by an air gap. The backflow preventer device and the piping downstream therefrom shall not be affected by carbon dioxide gas.

608.16.2 Connections to boilers. The potable supply to the boiler shall be equipped with a backflow preventer with an intermediate atmospheric vent complying with ASSE 1012 or CAN/CSA B64.3. Where conditioning chemicals are introduced into the system, the potable water connection shall be protected by an air gap or a reduced pressure principle backflow preventer, complying with ASSE 1013, CAN/CSA B64.4 or AWWA C511.

608.16.3 Heat exchangers. Heat exchangers utilizing an essentially toxic transfer fluid shall be separated from the potable water by double-wall construction. An air gap open to the atmosphere shall be provided between the two walls. Heat exchangers utilizing an essentially nontoxic transfer fluid shall be permitted to be of single-wall construction.

608.16.4 Connections to automatic fire sprinkler systems and standpipe systems. The potable water supply to automatic fire sprinkler and standpipe systems shall be protected against backflow by a double check-valve assembly or a reduced pressure principle backflow preventer.

Exceptions:

1. Where systems are installed as a portion of the water distribution system in accordance with the requirements of this code and are not provided with a fire department connection, isolation of the water supply system shall not be required.

2. Isolation of the water distribution system is not required for deluge, preaction or dry pipe systems.

608.16.4.1 Additives or nonpotable source. Where systems under continuous pressure contain chemical additives or antifreeze, or where systems are connected to a nonpotable secondary water supply, the potable water supply shall be protected against backflow by a reduced pressure principle backflow preventer. Where chemical additives or antifreeze are added to only a portion of an automatic fire sprinkler or standpipe system, the reduced pressure principle backflow preventer shall be permitted to be located so as to isolate that portion of the system. Where systems are not under continuous pressure, the potable water supply shall be protected against backflow by an air gap or a pipe applied atmospheric vacuum breaker conforming to ASSE 1001 or CAN/CSA B64.1.1.

608.16.5 Connections to lawn irrigation systems. The potable water supply to lawn irrigation systems shall be protected against backflow by an atmospheric-type vacuum breaker, a pressure-type vacuum breaker or a reduced pressure principle backflow preventer. A valve shall not be installed downstream from an atmospheric vacuum breaker. Where chemicals are introduced into the system, the potable water supply shall be protected against backflow by a reduced pressure principle backflow preventer.

608.16.6 Connections subject to backpressure. Where a potable water connection is made to a nonpotable line, fixture, tank, vat, pump or other equipment subject to back-pressure, the potable water connection shall be protected by a reduced pressure principle backflow preventer.

608.16.7 Chemical dispensers. Where chemical dispensers connect to the potable water distribution system, the water supply system shall be protected against backflow in accordance with Section 608.13.1, 608.13.2, 608.13.5, 608.13.6, 608.13.8 or 608.13.9.

608.16.8 Portable cleaning equipment. Where the portable cleaning equipment connects to the water distribution system, the water supply system shall be protected against backflow in accordance with Section 608.13.1, 608.13.2, 608.13.3, 608.13.7 or 608.13.8.

608.16.9 Dental pump equipment. Where dental pumping equipment connects to the water distribution system, the water supply system shall be protected against backflow in accordance with Section 608.13.1, 608.13.2, 608.13.5, 608.13.6 or 608.13.8.

608.17 Protection of individual water supplies. An individual water supply shall be located and constructed so as to be safeguarded against contamination in accordance with Sections 608.17.1 through 608.17.8.

608.17.1 Well locations. A potable ground water source or pump suction line shall not be located closer to potential sources of contamination than the distances shown in Table 608.17.1. In the event the underlying rock structure is limestone or fragmented shale, the local or state health department shall be consulted on well site location. The distances in Table 608.17.1 constitute minimum separation and shall be increased in areas of creviced rock or limestone, or where the direction of movement of the ground water is from sources of contamination toward the well.

608.17.2 Elevation. Well sites shall be positively drained and shall be at higher elevations than potential sources of contamination.

608.17.3 Depth. Private potable well supplies shall not be developed from a water table less than 10 feet (3048 mm) below the ground surface.

608.17.4 Water-tight casings. Each well shall be provided with a water-tight casing to a minimum distance of 10 feet (3048 mm) below the ground surface. All casings shall extend at least 6 inches (152 mm) above the well platform. The casing shall be large enough to permit installation of a separate drop pipe. Casings shall be sealed at the bottom in an impermeable stratum or extend several feet into the water-bearing stratum.

608.17.5 Drilled or driven well casings. Drilled or driven well casings shall be of steel or other approved material. Where drilled wells extend into a rock formation, the well casing shall extend to and set firmly in the formation. The annular space between the earth and the outside of the casing shall be filled with cement grout to a minimum distance of 10 feet (3048 mm) below the ground surface. In an instance of casing to rock installation, the grout shall extend to the rock surface.

TABLE 608.17.1
DISTANCE FROM CONTAMINATION TO
PRIVATE WATER SUPPLIES AND PUMP SUCTION LINES

SOURCE OF CONTAMINATION	DISTANCE (feet)
Barnyard	100
Farm silo	25
Pasture	100
Pumphouse floor drain of cast iron draining to ground surface	2
Seepage pits	50
Septic tank	25
Sewer	10
Subsurface disposal fields	50
Subsurface pits	50

For SI: 1 foot = 304.8 mm.

608.17.6 Dug or bored well casings. Dug or bored well casings shall be of water-tight concrete, tile, or galvanized or corrugated metal pipe to a minimum distance of 10 feet (3048 mm) below the ground surface. Where the water table is more than 10 feet (3048 mm) below the ground surface, the water-tight casing shall extend below the table surface. Well casings for dug wells or bored wells constructed with sections of concrete, tile, or galvanized or corrugated metal pipe shall be surrounded by 6 inches (152 mm) of grout poured into the hole between the outside of the casing and the ground to a minimum depth of 10 feet (3048 mm).

608.17.7 Cover. Every potable water well shall be equipped with an overlapping water-tight cover at the top of the well casing or pipe sleeve such that contaminated water or other substances are prevented from entering the well through the annular opening at the top of the well casing, wall or pipe sleeve. Covers shall extend downward at least 2 inches (51 mm) over the outside of the well casing or wall. A dug well cover shall be provided with a pipe sleeve permitting the withdrawal of the pump suction pipe, cylinder or jet body without disturbing the cover. Where pump sections or discharge pipes enter or leave a well through the side of the casing, the circle of contact shall be water tight.

608.17.8 Drainage. All potable water wells and springs shall be constructed such that surface drainage will be diverted away from the well or spring.

SECTION 609
HEALTH CARE PLUMBING

609.1 Scope. This section shall govern those aspects of health care plumbing systems that differ from plumbing systems in other structures. Health care plumbing systems shall conform to the requirements of this section in addition to the other requirements of this code. The provisions of this section shall apply to the special devices and equipment installed and maintained in the following occupancies: nursing homes, homes for the aged, orphanages, infirmaries, first aid stations, psychiatric facilities, clinics, professional offices of dentists and doctors, mortuaries, educational facilities, surgery, dentistry, research and testing laboratories, establishments manufacturing pharmaceutical drugs and medicines, and other structures with similar apparatus and equipment classified as plumbing.

609.2 Water service. All hospitals shall have two water service pipes installed in such a manner so as to minimize the potential for an interruption of the supply of water in the event of a water main or water service pipe failure.

609.3 Hot water. Hot water shall be provided to supply all of the hospital fixture, kitchen and laundry requirements. Special fixtures and equipment shall have hot water supplied at a temperature specified by the manufacturer. The hot water system shall be installed in accordance with Section 607.

609.4 Vacuum breaker installation. Vacuum breakers shall be installed a minimum of 6 inches (152 mm) above the flood level rim of the fixture or device in accordance with Section 608. The flood level rim of hose connections shall be the maximum height at which any hose is utilized.

609.5 Prohibited water closet and clinical sink supply. Jet- or water-supplied orifices, except those supplied by the flush connections, shall not be located in or connected with a water closet bowl or clinical sink. This section shall not prohibit an approved bidet installation.

609.6 Clinical, hydrotherapeutic and radiological equipment. All clinical, hydrotherapeutic, radiological or any equipment that is supplied with water or that discharges to the waste system shall conform to the requirements of this section and Section 608.

609.7 Condensate drain trap seal. A water supply shall be provided for cleaning, flushing and resealing the condensate trap, and the trap shall discharge through an air gap in accordance with Section 608.

609.8 Valve leakage diverter. Each water sterilizer filled with water through directly connected piping shall be equipped with an approved leakage diverter or bleed line on the water supply control valve to indicate and conduct any leakage of unsterile water away from the sterile zone.

SECTION 610
DISINFECTION OF POTABLE WATER SYSTEM

610.1 General. New or repaired potable water systems shall be purged of deleterious matter and disinfected prior to utilization. The method to be followed shall be that prescribed by the health authority or water purveyor having jurisdiction or, in the absence of a prescribed method, the procedure described in either AWWA C651 or AWWA C652, or as described in this section. This requirement shall apply to "on-site" or "in-plant" fabrication of a system or to a modular portion of a system.

1. The pipe system shall be flushed with clean, potable water until dirty water does not appear at the points of outlet.

2. The system or part thereof shall be filled with a water/chlorine solution containing at least 50 parts per million (50 mg/L) of chlorine, and the system or part thereof shall be valved off and allowed to stand for 24 hours; or the system

or part thereof shall be filled with a water/chlorine solution containing at least 200 parts per million (200 mg/L) of chlorine and allowed to stand for 3 hours.

3. Following the required standing time, the system shall be flushed with clean potable water until the chlorine is purged from the system.

4. The procedure shall be repeated where shown by a bacteriological examination that contamination remains present in the system.

SECTION 611
DRINKING WATER TREATMENT UNITS

611.1 Design. Drinking water treatment units shall meet the requirements of NSF 42, NSF 44, NSF 53 or NSF 62.

611.2 Reverse osmosis systems. The discharge from a reverse osmosis drinking water treatment unit shall enter the drainage system through an air gap or an air gap device that meets the requirements of NSF 58.

611.3 Connection tubing. The tubing to and from drinking water treatment units shall be of a size and material as recommended by the manufacturer. The tubing shall comply with NSF 14, NSF 42, NSF 44, NSF 53, NSF 58 or NSF 61.

SECTION 612
SOLAR SYSTEMS

612.1 Solar systems. The construction, installation, alterations and repair of systems, equipment and appliances intended to utilize solar energy for space heating or cooling, domestic hot water heating, swimming pool heating or process heating shall be in accordance with the *International Mechanical Code*.

SECTION 613
TEMPERATURE CONTROL
DEVICES AND VALVES

613.1 Temperature-actuated mixing valves. Temperature-actuated mixing valves, which are installed to reduce water temperatures to defined limits, shall comply with ASSE 1017.

CHAPTER 7
SANITARY DRAINAGE

SECTION 701
GENERAL

701.1 Scope. The provisions of this chapter shall govern the materials, design, construction and installation of sanitary drainage systems.

701.2 Sewer required. Every building in which plumbing fixtures are installed and all premises having drainage piping shall be connected to a public sewer, where available, or an approved private sewage disposal system in accordance with the *International Private Sewage Disposal Code.*

701.3 Separate sewer connection. Every building having plumbing fixtures installed and intended for human habitation, occupancy or use on premises abutting on a street, alley or easement in which there is a public sewer shall have a separate connection with the sewer. Where located on the same lot, multiple buildings shall not be prohibited from connecting to a common building sewer that connects to the public sewer.

701.4 Sewage treatment. Sewage or other waste from a plumbing system that is deleterious to surface or subsurface waters shall not be discharged into the ground or into any waterway unless it has first been rendered innocuous through subjection to an approved form of treatment.

701.5 Damage to drainage system or public sewer. Wastes detrimental to the public sewer system or to the functioning of the sewage-treatment plant shall be treated and disposed of in accordance with Section 1003 as directed by the code official.

701.6 Tests. The sanitary drainage system shall be tested in accordance with Section 312.

701.7 Connections. Direct connection of a steam exhaust, blowoff or drip pipe shall not be made with the building drainage system. Wastewater when discharged into the building drainage system shall be at a temperature not higher than 140°F (60°C). When higher temperatures exist, approved cooling methods shall be provided.

701.8 Engineered systems. Engineered sanitary drainage systems shall conform to the provisions of Sections 105.4 and 714.

701.9 Drainage piping in food service areas. Exposed soil or waste piping shall not be installed above any working, storage or eating surfaces in food service establishments.

SECTION 702
MATERIALS

702.1 Above-ground sanitary drainage and vent pipe. Above-ground soil, waste and vent pipe shall conform to one of the standards listed in Table 702.1.

702.2 Underground building sanitary drainage and vent pipe. Underground building sanitary drainage and vent pipe shall conform to one of the standards listed in Table 702.2.

702.3 Building sewer pipe. Building sewer pipe shall conform to one of the standards listed in Table 702.3.

702.4 Fittings. Pipe fittings shall be approved for installation with the piping material installed and shall conform to the respective pipe standards or one of the standards listed in Table 702.4.

702.5 Chemical waste system. A chemical waste system shall be completely separated from the sanitary drainage system. The chemical waste shall be treated in accordance with Section 803.2 before discharging to the sanitary drainage system. Separate drainage systems for chemical wastes and vent pipes shall be of an approved material that is resistant to corrosion and degradation for the concentrations of chemicals involved.

702.6 Lead bends and traps. Lead bends and traps shall not be less than 0.125 inch (3.2mm) wall thickness.

TABLE 702.1
ABOVE-GROUND DRAINAGE AND VENT PIPE

MATERIAL	STANDARD
Acrylonitrile butadiene styrene (ABS) plastic pipe	ASTM D 2661; ASTM F 628; CSA B181.1
Brass pipe	ASTM B 43
Cast-iron pipe	ASTM A 74; CISPI 301; ASTM A 888
Coextruded composite ABS DWV schedule 40 IPS pipe (solid)	ASTM F 1488
Coextruded composite ABS DWV schedule 40 IPS pipe (cellular core)	ASTM F 1488
Coextruded composite PVC DWV schedule 40 IPS pipe (solid)	ASTM F 1488
Coextruded composite PVC DWV schedule 40 IPS pipe (cellular core)	ASTM F 1488
Coextruded composite PVC IPS-DR, PS140, PS200 DWV	ASTM F 1488
Copper or copper-alloy pipe	ASTM B 42; ASTM B 302
Copper or copper-alloy tubing (Type K, L, M or DWV)	ASTM B 75; ASTM B 88; ASTM B 251; ASTM B 306
Galvanized steel pipe	ASTM A 53
Glass pipe	ASTM C 1053
Polyolefin pipe	CAN/CSA-B181.3
Polyvinyl chloride (PVC) plastic pipe (Type DWV)	ASTM D 2665; ASTM D 2949; ASTM F 891; CSA B181.2; ASTM F 1488
Stainless steel drainage systems, Types 304 and 316L	ASME A112.3.1

TABLE 702.2
UNDERGROUND BUILDING DRAINAGE AND VENT PIPE

MATERIAL	STANDARD
Acrylonitrile butadiene styrene (ABS) plastic pipe	ASTM D 2661; ASTM F 628; CSA B181.1
Asbestos-cement pipe	ASTM C 428
Cast-iron pipe	ASTM A 74; CISPI 301; ASTM A 888
Coextruded composite ABS DWV schedule 40 IPS pipe (solid)	ASTM F 1488
Coextruded composite ABS DWV schedule 40 IPS pipe (cellular core)	ASTM F 1488
Coextruded composite PVC DWV schedule 40 IPS pipe (solid)	ASTM F 1488
Coextruded compsoite PVC DWV schedule 40 IPS pipe (cellular core)	ASTM F 1488
Coextruded composite PVC IPS-DR, PS140, PS200 DWV	ASTM F 1488
Copper or copper alloy tubing (Type K, L, M or DWV)	ASTM B 75; ASTM B 88; ASTM B 251; ASTM B 306
Polyolefin pipe	CAN/CSA-B181.3
Polyvinyl chloride (PVC) plastic pipe (Type DWV)	ASTM D 2665; ASTM D 2949; ASTM F 891; CSA-B181.2
Stainless steel drainage systems, Type 316L	ASME A112.3.1

TABLE 702.3
BUILDING SEWER PIPE

MATERIAL	STANDARD
Acrylonitrile butadiene styrene (ABS) plastic pipe	ASTM D 2661; ASTM D 2751; CSA F 628
Asbestos-cement	ASTM C 428
Cast-iron pipe	ASTM A 74; ASTM A 888; CISPI 301
Coextruded composite ABS DWV schedule 40 IPS pipe (solid)	ASTM F 1488
Coextruded composite ABS DWV schedule 40 IPS pipe (cellular core)	ASTM F 1488
Coextruded composite PVC DWV schedule 40 IPS pipe (solid)	ASTM F 1488
Coextruded composite PVC DWV schedule 40 IPS pipe (cellular core)	ASTM F 1488
Coextruded composite PVC IPS-DR, PS140, PS200, DWV	ASTM F 1488
Coextruded composite ABS sewer and drain DR-PS in PS35, PS50, PS100, PS140, PS200	ASTM F 1488
Coextruded composite PVC sewer and drain DR-PS in PS35, PS50, PS100, PS140, PS200	ASTM F 1488
Concrete pipe	ASTM C14; ASTM C76; CAN/CSA A257.1M; CAN/CSA A257.2M
Copper or copper-alloy tubing (Type K or L)	ASTM B 75; ASTM B 88; ASTM B 251
Polyethylene (PE) plastic pipe (SDR-PR)	ASTM F 714
Polyvinyl chloride (PVC) plastic pipe (Type DWV, SDR26, SDR35, SDR41, PS50 or PS100)	ASTM D 2665; ASTM D 2949; ASTM D 3034; ASTM F 891; CSA B182.2; CAN/CSA B182.4
Stainless steel drainage systems, types 304 and 316L	ASME A112.3.1
Vitrified clay pipe	ASTM C 4; ASTM C 700

TABLE 702.4
PIPE FITTINGS

MATERIAL	STANDARD
Acrylonitrile butadiene styrene (ABS) plastic pipe	ASTM D 3311; CSA B181.1; ASTM D 2661
Cast iron	ASME B 16.4; ASME B 16.12; ASTM A 74; ASTM A 888; CISPI 301
Coextruded composite ABS DWV schedule 40 IPS pipe (solid or cellular core)	ASTM D 2661; ASTM D 3311; ASTM F 628
Coextruded composite PVC DWV schedule 40 IPS-DR, PS140, PS200 (solid or cellular core)	ASTM D 2665; ASTM D 3311; ASTM F 891
Coextruded composite ABS sewer and drain DR-PS in PS35, PS50, PS100, PS140, PS200	ASTM D 2751
Coextruded composite PVC sewer and drain DR-PS in PS35, PS50, PS100, PS140, PS200	ASTM D 3034
Copper or copper alloy	ASME B 16.15; ASME B 16.18; ASME B 16.22; ASME B 16.23; ASME B 16.26; ASME B 16.29
Glass	ASTM C 1053
Gray iron and ductile iron	AWWA C 110
Malleable iron	ASME B 16.3
Polyvinyl chloride (PVC) plastic	ASTM D 3311; ASTM D 2665; ASTM F 1866
Stainless steel drainage systems, Types 304 and 316L	ASME A 112.3.1
Steel	ASME B 16.9; ASME B16.11; ASME B16.28

SECTION 703
BUILDING SEWER

703.1 Building sewer pipe near the water service. Where the building sewer is installed within 5 feet (1524 mm) of the water service, as provided for in Section 603.2, the building sewer pipe shall conform to one of the standards for ABS plastic pipe, cast-iron pipe, copper or copper-alloy tubing, or PVC plastic pipe listed in Table 702.3.

703.2 Drainage pipe in filled ground. Where a building sewer or building drain is installed on filled or unstable ground, the drainage pipe shall conform to one of the standards for ABS plastic pipe, cast-iron pipe, copper or copper-alloy tubing, or PVC plastic pipe listed in Table 702.3.

703.3 Sanitary and storm sewers. Where separate systems of sanitary drainage and storm drainage are installed in the same property, the sanitary and storm building sewers or drains shall be permitted to be laid side by side in one trench.

703.4 Existing building sewers and drains. Existing building sewers and drains shall connect with new building sewer and drainage systems only where found by examination and test to conform to the new system in quality of material. The code offi-

cial shall notify the owner to make the changes necessary to conform to this code.

703.5 Cleanouts on building sewers. Cleanouts on building sewers shall be located as set forth in Section 708.

SECTION 704
DRAINAGE PIPING INSTALLATION

704.1 Slope of horizontal drainage piping. Horizontal drainage piping shall be installed in uniform alignment at uniform slopes. The minimum slope of a horizontal drainage pipe shall be in accordance with Table 704.1.

TABLE 704.1
SLOPE OF HORIZONTAL DRAINAGE PIPE

SIZE (inches)	MINIMUM SLOPE (inch per foot)
$2^1/_2$ or less	$^1/_4$
3 to 6	$^1/_8$
8 or larger	$^1/_{16}$

For SI: 1 inch = 25.4 mm, 1 inch per foot = 0.083.3 mm/m.

704.2 Change in size. The size of the drainage piping shall not be reduced in size in the direction of the flow. A 4-inch by 3-inch (102 mm by 76 mm) water closet connection shall not be considered as a reduction in size.

704.3 Connections to offsets and bases of stacks. Horizontal branches shall connect to the bases of stacks at a point located not less than 10 times the diameter of the drainage stack downstream from the stack. Except as prohibited by Section 711.2, horizontal branches shall connect to horizontal stack offsets at a point located not less than 10 times the diameter of the drainage stack downstream from the upper stack.

704.4 Future fixtures. Drainage piping for future fixtures shall terminate with an approved cap or plug.

704.5 Dead ends. In the installation or removal of any part of a drainage system, dead ends shall be prohibited. Cleanout extensions and approved future fixture drainage piping shall not be considered as dead ends.

SECTION 705
JOINTS

705.1 General. This section contains provisions applicable to joints specific to sanitary drainage piping.

705.2 ABS plastic. Joints between ABS plastic pipe or fittings shall comply with Sections 705.2.1 through 705.2.3.

    705.2.1 Mechanical joints. Mechanical joints on drainage pipes shall be made with an elastomeric seal conforming to ASTM C 1173, ASTM D 3212 or CAN/CSA-B602. Mechanical joints shall be installed only in underground systems unless otherwise approved. Joints shall be installed in accordance with the manufacturer's instructions.

    705.2.2 Solvent cementing. Joint surfaces shall be clean and free from moisture. Solvent cement that conforms to ASTM D 2235 or CSA B181.1 shall be applied to all joint surfaces. The joint shall be made while the cement is wet.

Joints shall be made in accordance with ASTM D 2235, ASTM D 2661, ASTM F 628 or CSA B181.1. Solvent-cement joints shall be permitted above or below ground.

705.2.3 Threaded joints. Threads shall conform to ASME B1.20.1. Schedule 80 or heavier pipe shall be permitted to be threaded with dies specifically designed for plastic pipe. Approved thread lubricant or tape shall be applied on the male threads only.

705.3 Asbestos-cement. Joints between asbestos-cement pipe or fittings shall be made with a sleeve coupling of the same composition as the pipe, sealed with an elastomeric ring conforming to ASTM D 1869.

705.4 Brass. Joints between brass pipe or fittings shall comply with Sections 705.4.1 through 705.4.4.

705.4.1 Brazed joints. All joint surfaces shall be cleaned. An approved flux shall be applied where required. The joint shall be brazed with a filler metal conforming to AWS A5.8.

705.4.2 Mechanical joints. Mechanical joints shall be installed in accordance with the manufacturer's instructions.

705.4.3 Threaded joints. Threads shall conform to ASME B1.20.1. Pipe-joint compound or tape shall be applied on the male threads only.

705.4.4 Welded joints. All joint surfaces shall be cleaned. The joint shall be welded with an approved filler metal.

705.5 Cast iron. Joints between cast-iron pipe or fittings shall comply with Sections 705.5.1 through 705.5.3.

705.5.1 Caulked joints. Joints for hub and spigot pipe shall be firmly packed with oakum or hemp. Molten lead shall be poured in one operation to a depth of not less than 1 inch (25 mm). The lead shall not recede more than 0.125 inch (3.2 mm) below the rim of the hub and shall be caulked tight. Paint, varnish or other coatings shall not be permitted on the jointing material until after the joint has been tested and approved. Lead shall be run in one pouring and shall be caulked tight. Acid-resistant rope and acidproof cement shall be permitted.

705.5.2 Compression gasket joints. Compression gaskets for hub and spigot pipe and fittings shall conform to ASTM C 564. Gaskets shall be compressed when the pipe is fully inserted.

705.5.3 Mechanical joint coupling. Mechanical joint couplings for hubless pipe and fittings shall comply with CISPI 310 or ASTM C 1277. The elastomeric sealing sleeve shall conform to ASTM C 564 or CAN/CSA B602 and shall be provided with a center stop. Mechanical joint couplings shall be installed in accordance with the manufacturer's installation instructions.

705.6 Concrete joints. Joints between concrete pipe and fittings shall be made with an elastomeric seal conforming to ASTM C 443, ASTM C 1173, CAN/CSA A257.3M or CAN/CSA-B602.

705.7 Coextruded composite ABS pipe, joints. Joints between coextruded composite pipe with an ABS outer layer or ABS fittings shall comply with Sections 705.7.1 and 705.7.2.

705.7.1 Mechanical joints. Mechanical joints on drainage pipe shall be made with an elastomeric seal conforming to ASTM C1173, ASTM D 3212 or CAN/CSA B602. Mechanical joints shall not be installed in above-ground systems, unless otherwise approved. Joints shall be installed in accordance with the manufacturer's instructions.

705.7.2 Solvent cementing. Joint surfaces shall be clean and free from moisture. Solvent cement that conforms to ASTM D 2235 or CSA-B181.1 shall be applied to all joint surfaces. The joint shall be made while the cement is wet. Joints shall be made in accordance with ASTM D 2235, ASTM D 2661, ASTM F 628 or CSA B181.1. Solvent-cement joints shall be permitted above or below ground.

705.8 Coextruded composite PVC pipe. Joints between coextruded composite pipe with a PVC outer layer or PVC fittings shall comply with Sections 705.8.1 and 705.8.2.

705.8.1 Mechanical joints. Mechanical joints on drainage pipe shall be made with an elastomeric seal conforming to ASTM D 3212. Mechanical joints shall not be installed in above-ground systems, unless otherwise approved. Joints shall be installed in accordance with the manufacturer's instructions.

705.8.2 Solvent cementing. Joint surfaces shall be clean and free from moisture. A purple primer that conforms to ASTM F 656 shall be applied. Solvent cement not purple in color and conforming to ASTM D 2564, CSA B137.3, CSA B181.2 or CSA B182.1 shall be applied to all joint surfaces. The joint shall be made while the cement is wet and shall be in accordance with ASTM D 2855. Solvent-cement joints shall be permitted above or below ground.

705.9 Copper pipe. Joints between copper or copper-alloy pipe or fittings shall comply with Sections 705.9.1 through 705.9.5.

705.9.1 Brazed joints. All joint surfaces shall be cleaned. An approved flux shall be applied where required. The joint shall be brazed with a filler metal conforming to AWS A5.8.

705.9.2 Mechanical joints. Mechanical joints shall be installed in accordance with the manufacturer's instructions.

705.9.3 Soldered joints. Solder joints shall be made in accordance with the methods of ASTM B 828. All cut tube ends shall be reamed to the full inside diameter of the tube end. All joint surfaces shall be cleaned. A flux conforming to ASTM B 813 shall be applied. The joint shall be soldered with a solder conforming to ASTM B 32.

705.9.4 Threaded joints. Threads shall conform to ASME B1.20.1. Pipe-joint compound or tape shall be applied on the male threads only.

705.9.5 Welded joints. All joint surfaces shall be cleaned. The joint shall be welded with an approved filler metal.

705.10 Copper tubing. Joints between copper or copper-alloy tubing or fittings shall comply with Sections 705.10.1 through 705.10.3.

705.10.1 Brazed joints. All joint surfaces shall be cleaned. An approved flux shall be applied where required. The joint shall be brazed with a filler metal conforming to AWS A5.8.

705.10.2 Mechanical joints. Mechanical joints shall be installed in accordance with the manufacturer's instructions.

705.10.3 Soldered joints. Solder joints shall be made in accordance with the methods of ASTM B 828. All cut tube ends shall be reamed to the full inside diameter of the tube end. All joint surfaces shall be cleaned. A flux conforming to ASTM B 813 shall be applied. The joint shall be soldered with a solder conforming to ASTM B 32.

705.11 Borosilicate glass joints. Glass-to-glass connections shall be made with a bolted compression-type stainless steel (300 series) coupling with contoured acid-resistant elastomeric compression ring and a fluorocarbon polymer inner seal ring; or with caulked joints in accordance with Section 705.11.1.

705.11.1 Caulked joints. Every lead-caulked joint for hub and spigot soil pipe shall be firmly packed with oakum or hemp and filled with molten lead not less than 1 inch (25 mm) deep and not to extend more than 0.125 inch (3.2 mm) below the rim of the hub. Paint, varnish or other coatings shall not be permitted on the jointing material until after the joint has been tested and approved. Lead shall be run in one pouring and shall be caulked tight. Acid-resistant rope and acidproof cement shall be permitted.

705.12 Steel. Joints between galvanized steel pipe or fittings shall comply with Sections 705.12.1 and 705.12.2.

705.12.1 Threaded joints. Threads shall conform to ASME B1.20.1. Pipe-joint compound or tape shall be applied on the male threads only.

705.12.2 Mechanical joints. Joints shall be made with an approved elastomeric seal. Mechanical joints shall be installed in accordance with the manufacturer's instructions.

705.13 Lead. Joints between lead pipe or fittings shall comply with Sections 705.13.1 and 705.13.2.

705.13.1 Burned. Burned joints shall be uniformly fused together into one continuous piece. The thickness of the joint shall be at least as thick as the lead being joined. The filler metal shall be of the same material as the pipe.

705.13.2 Wiped. Joints shall be fully wiped, with an exposed surface on each side of the joint not less than 0.75 inch (19.1 mm). The joint shall be at least 0.325 inch (9.5 mm) thick at the thickest point.

705.14 PVC plastic. Joints between PVC plastic pipe or fittings shall comply with Sections 705.14.1 through 705.14.3.

705.14.1 Mechanical joints. Mechanical joints on drainage pipe shall be made with an elastomeric seal conforming to ASTM C 1173, ASTM D 3212 or CAN/CSA-B602. Mechanical joints shall not be installed in above-ground systems, unless otherwise approved. Joints shall be installed in accordance with the manufacturer's instructions.

705.14.2 Solvent cementing. Joint surfaces shall be clean and free from moisture. A purple primer that conforms to ASTM F 656 shall be applied. Solvent cement not purple in color and conforming to ASTM D 2564, CSA B137.3, CSA B181.2 or CSA B182.1 shall be applied to all joint surfaces. The joint shall be made while the cement is wet and shall be in accordance with ASTM D 2855. Solvent-cement joints shall be permitted above or below ground.

705.14.3 Threaded joints. Threads shall conform to ASME B1.20.1. Schedule 80 or heavier pipe shall be permitted to be threaded with dies specifically designed for plastic pipe. Approved thread lubricant or tape shall be applied on the male threads only.

705.15 Vitrified clay. Joints between vitrified clay pipe or fittings shall be made with an elastomeric seal conforming to ASTM C 425, ASTM C 1173 or CAN/CSA-B602.

705.16 Joints between different materials. Joints between different piping materials shall be made with a mechanical joint of the compression or mechanical-sealing type conforming to ASTM C 1173, ASTM C 1460 or ASTM C 1461. Connectors and adapters shall be approved for the application and such joints shall have an elastomeric seal conforming to ASTM C 425, ASTM C 443, ASTM C 564, ASTM C 1440, ASTM D 1869, ASTM F 477, CAN/CSA A257.3M orCAN/CSA B602, or as required in Sections 705.16.1 through 705.16.5. Joints between glass pipe and other types of materials shall be made with adapters having a TFE seal. Joints shall be installed in accordance with the manufacturer's instructions.

705.16.1 Copper or copper-alloy tubing to cast-iron hub pipe. Joints between copper or copper-alloy tubing and cast-iron hub pipe shall be made with a brass ferrule or compression joint. The copper or copper-alloy tubing shall be soldered to the ferrule in an approved manner, and the ferrule shall be joined to the cast-iron hub by a caulked joint or a mechanical compression joint.

705.16.2 Copper or copper-alloy tubing to galvanized steel pipe. Joints between copper or copper-alloy tubing and galvanized steel pipe shall be made with a brass converter fitting or dielectric fitting. The copper tubing shall be soldered to the fitting in an approved manner, and the fitting shall be screwed to the threaded pipe.

705.16.3 Cast-iron pipe to galvanized steel or brass pipe. Joints between cast-iron and galvanized steel or brass pipe shall be made by either caulked or threaded joints or with an approved adapter fitting.

705.16.4 Plastic pipe or tubing to other piping material. Joints between different grades of plastic pipe or between plastic pipe and other piping material shall be made with an approved adapter fitting. Joints between plastic pipe and cast-iron hub pipe shall be made by a caulked joint or a mechanical compression joint.

705.16.5 Lead pipe to other piping material. Joints between lead pipe and other piping material shall be made by a wiped joint to a caulking ferrule, soldering nipple, or bushing or shall be made with an approved adapter fitting.

705.16.6 Borosilicate glass to other materials. Joints between glass pipe and other types of materials shall be made with adapters having a TFE seal and shall be installed in accordance with the manufacturer's instructions.

705.16.7 Stainless steel drainage systems to other materials. Joints between stainless steel drainage systems and other piping materials shall be made with approved mechanical couplings.

705.17 Drainage slip joints. Slip joints shall comply with Section 405.8.

705.18 Caulking ferrules. Ferrules shall be of red brass and shall be in accordance with Table 705.18.

TABLE 705.18
CAULKING FERRULE SPECIFICATIONS

PIPE SIZES (inches)	INSIDE DIAMETER (inches)	LENGTH (inches)	MINIMUM WEIGHT EACH
2	$2^1/_4$	$4^1/_2$	1 pound
3	$3^1/_4$	$4^1/_2$	1 pound 12 ounces
4	$4^1/_4$	$4^1/_2$	2 pounds 8 ounces

For SI: 1 inch = 25.4 mm, 1 ounce = 28.35 g, 1 pound = 0.454 kg.

705.19 Soldering bushings. Soldering bushings shall be of red brass and shall be in accordance with Table 705.19.

TABLE 705.19
SOLDERING BUSHING SPECIFICATIONS

PIPE SIZES (inches)	MINIMUM WEIGHT EACH
$1^1/_4$	6 ounces
$1^1/_2$	8 ounces
2	14 ounces
$2^1/_2$	1 pound 6 ounces
3	2 pounds
4	3 pounds 8 ounces

For SI: 1 inch = 25.4 mm, 1 ounce = 28.35 g, 1 pound = 0.454 kg.

705.20 Stainless steel drainage systems. O-ring joints for stainless steel drainage systems shall be made with an approved elastomeric seal.

SECTION 706
CONNECTIONS BETWEEN DRAINAGE PIPING AND FITTINGS

706.1 Connections and changes in direction. All connections and changes in direction of the sanitary drainage system shall be made with approved drainage fittings. Connections between drainage piping and fixtures shall conform to Section 405.

706.2 Obstructions. The fittings shall not have ledges, shoulders or reductions capable of retarding or obstructing flow in the piping. Threaded drainage pipe fittings shall be of the recessed drainage type.

706.3 Installation of fittings. Fittings shall be installed to guide sewage and waste in the direction of flow. Change in direction shall be made by fittings installed in accordance with Table 706.3. Change in direction by combination fittings, side inlets or increasers shall be installed in accordance with Table 706.3 based on the pattern of flow created by the fitting. Double sanitary tee patterns shall not receive the discharge of back-to-back water closets and fixtures or appliances with pumping action discharge.

Exception: Back-to-back water closet connections to double sanitary tees shall be permitted where the horizontal developed length between the outlet of the water closet and the connection to the double sanitary tee pattern is 18 inches (457 mm) or greater.

TABLE 706.3
FITTINGS FOR CHANGE IN DIRECTION

TYPE OF FITTING PATTERN	CHANGE IN DIRECTION		
	Horizontal to vertical	Vertical to horizontal	Horizontal to horizontal
Sixteenth bend	X	X	X
Eighth bend	X	X	X
Sixth bend	X	X	X
Quarter bend	X	X[a]	X[a]
Short sweep	X	X[a,b]	X[a]
Long sweep	X	X	X
Sanitary tee	X[c]	—	—
Wye	X	X	X
Combination wye and eighth bend	X	X	X

For SI: 1 inch = 25.4 mm.

a. The fittings shall only be permitted for a 2-inch or smaller fixture drain.

b. Three inches or larger.

c. For a limitation on double sanitary tees, see Section 706.3.

SECTION 707
PROHIBITED JOINTS AND CONNECTIONS

707.1 Prohibited joints. The following types of joints and connections shall be prohibited:

1. Cement or concrete joints.

2. Mastic or hot-pour bituminous joints.

3. Joints made with fittings not approved for the specific installation.

4. Joints between different diameter pipes made with elastomeric rolling O-rings.

5. Solvent-cement joints between different types of plastic pipe.

6. Saddle-type fittings.

SECTION 708
CLEANOUTS

708.1 Scope. This section shall govern the size, location, installation and maintenance of drainage pipe cleanouts.

708.2 Cleanout plugs. Cleanout plugs shall be brass or plastic, or other approved materials. Brass cleanout plugs shall be utilized with metallic drain, waste and vent piping only, and shall conform to ASTM A 74, ASME A112.3.1 or ASME

A112.36.2M. Cleanouts with plate-style access covers shall be fitted with corrosion-resisting fasteners. Plastic cleanout plugs shall conform to the requirements of Section 702.4. Plugs shall have raised square or countersunk square heads. Countersunk heads shall be installed where raised heads are a trip hazard. Cleanout plugs with borosilicate glass systems shall be of borosilicate glass.

708.3 Where required. Cleanouts shall be located in accordance with Sections 708.3.1 through 708.3.5.

708.3.1 Horizontal drains within buildings. All horizontal drains shall be provided with cleanouts located not more than 100 feet (30 480 mm) apart.

708.3.2 Building sewers. Building sewers shall be provided with cleanouts located not more than 100 feet (30 480 mm) apart measured from the upstream entrance of the cleanout. For building sewers 8 inches (203 mm) and larger, manholes shall be provided and located not more than 200 feet (60 960 mm) from the junction of the building drain and building sewer, at each change in direction and at intervals of not more than 400 feet (122 m) apart. Manholes and manhole covers shall be of an approved type.

708.3.3 Changes of direction. Cleanouts shall be installed at each change of direction of the building drain or horizontal waste or soil lines greater than 45 degrees (0.79 rad). Where more than one change of direction occurs in a run of piping, only one cleanout shall be required for each 40 feet (12 192 mm) of developed length of the drainage piping.

708.3.4 Base of stack. A cleanout shall be provided at the base of each waste or soil stack.

708.3.5 Building drain and building sewer junction. There shall be a cleanout near the junction of the building drain and the building sewer. The cleanout shall be either inside or outside the building wall and shall be brought up to the finished ground level or to the basement floor level. An approved two-way cleanout is allowed to be used at this location to serve as a required cleanout for both the building drain and building sewer. The cleanout at the junction of the building drain and building sewer shall not be required if the cleanout on a 3-inch (76 mm) or larger diameter soil stack is located within a developed length of 10 feet (3048 mm) of the building drain and building sewer connection. The minimum size of the cleanout at the junction of the building drain and building sewer shall comply with Section 708.7.

708.3.6 Manholes. Manholes serving a building drain shall have secured gas-tight covers and shall be located in accordance with Section 708.3.2.

708.4 Concealed piping. Cleanouts on concealed piping or piping under a floor slab or in a crawl space of less than 24 inches (610 mm) in height or a plenum shall be extended through and terminate flush with the finished wall, floor or ground surface or shall be extended to the outside of the building. Cleanout plugs shall not be covered with cement, plaster or any other permanent finish material. Where it is necessary to conceal a cleanout or to terminate a cleanout in an area subject to vehicular traffic, the covering plate, access door or cleanout shall be of an approved type designed and installed for this purpose.

708.5 Opening direction. Every cleanout shall be installed to open to allow cleaning in the direction of the flow of the drainage pipe or at right angles thereto.

708.6 Prohibited installation. Cleanout openings shall not be utilized for the installation of new fixtures, except where approved and where another cleanout of equal access and capacity is provided.

708.7 Minimum size. Cleanouts shall be the same nominal size as the pipe they serve up to 4 inches (102 mm). For pipes larger than 4 inches (102 mm) nominal size, the minimum size of the cleanout shall be 4 inches (102 mm).

Exceptions:

1. "P" trap connections with slip joints or ground joint connections, or stack cleanouts that are not more than one pipe diameter smaller than the drain served, shall be permitted.

2. Cast-iron cleanout sizing shall be in accordance with referenced standards in Table 702.4, ASTM A 74 for hub and spigot fittings or ASTM A 888 or CISPI 301 for hubless fittings.

708.8 Clearances. Cleanouts on 6-inch (153 mm) and smaller pipes shall be provided with a clearance of not less than 18 inches (457 mm) for rodding. Cleanouts on 8-inch (203 mm) and larger pipes shall be provided with a clearance of not less than 36 inches (914 mm) for rodding.

708.9 Access. Access shall be provided to all cleanouts.

SECTION 709
FIXTURE UNITS

709.1 Values for fixtures. Drainage fixture unit values as given in Table 709.1 designate the relative load weight of different kinds of fixtures that shall be employed in estimating the total load carried by a soil or waste pipe, and shall be used in connection with Tables 710.1(1) and 710.1(2) of sizes for soil, waste and vent pipes for which the permissible load is given in terms of fixture units.

709.2 Fixtures not listed in Table 709.1. Fixtures not listed in Table 709.1 shall have a drainage fixture unit load based on the outlet size of the fixture in accordance with Table 709.2. The minimum trap size for unlisted fixtures shall be the size of the drainage outlet but not less than 1.25 inches (32 mm).

709.3 Values for continuous and semicontinuous flow. Drainage fixture unit values for continuous and semicontinuous flow into a drainage system shall be computed on the basis that 1 gpm (0.06 L/s) of flow is equivalent to two fixture units.

709.4 Values for indirect waste receptor. The drainage fixture unit load of an indirect waste receptor receiving the discharge of indirectly connected fixtures shall be the sum of the drainage fixture unit values of the fixtures that discharge to the receptor, but not less than the drainage fixture unit value given for the indirect waste receptor in Table 709.1 or 709.2.

TABLE 709.1
DRAINAGE FIXTURE UNITS FOR FIXTURES AND GROUPS

FIXTURE TYPE	DRAINAGE FIXTURE UNIT VALUE AS LOAD FACTORS	MINIMUM SIZE OF TRAP (inches)
Automatic clothes washers, commercial[a,g]	3	2
Automatic clothes washers, residential[g]	2	2
Bathroom group as defined in Section 202 (1.6 gpf water closet)[f]	5	—
Bathroom group as defined in Section 202 (water closet flushing greater than 1.6 gpf)[f]	6	—
Bathtub[b] (with or without overhead shower or whirlpool attachments)	2	$1^1/_2$
Bidet	1	$1^1/_4$
Combination sink and tray	2	$1^1/_2$
Dental lavatory	1	$1^1/_4$
Dental unit or cuspidor	1	$1^1/_4$
Dishwashing machine,[c] domestic	2	$1^1/_2$
Drinking fountain	$^1/_2$	$1^1/_4$
Emergency floor drain	0	2
Floor drains	2	2
Kitchen sink, domestic	2	$1^1/_2$
Kitchen sink, domestic with food waste grinder and/or dishwasher	2	$1^1/_2$
Laundry tray (1 or 2 compartments)	2	$1^1/_2$
Lavatory	1	$1^1/_4$
Shower	2	$1^1/_2$
Sink	2	$1^1/_2$
Urinal	4	Note d
Urinal, 1 gallon per flush or less	2[e]	Note d
Wash sink (circular or multiple) each set of faucets	2	$1^1/_2$
Water closet, flushometer tank, public or private	4[e]	Note d
Water closet, private (1.6 gpf)	3[e]	Note d
Water closet, private (flushing greater than 1.6 gpf)	4[e]	Note d
Water closet, public (1.6 gpf)	4[e]	Note d
Water closet, public (flushing greater than 1.6 gpf)	6[e]	Note d

For SI: 1 inch = 25.4 mm, 1 gallon = 3.785 L.

a. For traps larger than 3 inches, use Table 709.2.

b. A showerhead over a bathtub or whirlpool bathtub attachment does not increase the drainage fixture unit value.

c. See Sections 709.2 through 709.4 for methods of computing unit value of fixtures not listed in this table or for rating of devices with intermittent flows.

d. Trap size shall be consistent with the fixture outlet size.

e. For the purpose of computing loads on building drains and sewers, water closets and urinals shall not be rated at a lower drainage fixture unit unless the lower values are confirmed by testing.

f. For fixtures added to a dwelling unit bathroom group, add the DFU value of those additional fixtures to the bathroom group fixture count.

g. See Section 406.3 for sizing requirements for fixture drain, branch drain, and drainage stack for an automatic clothes washer standpipe.

TABLE 709.2
DRAINAGE FIXTURE UNITS FOR FIXTURE DRAINS OR TRAPS

FIXTURE DRAIN OR TRAP SIZE (inches)	DRAINAGE FIXTURE UNIT VALUE
1¼	1
1½	2
2	3
2½	4
3	5
4	6

For SI: 1 inch = 25.4 mm.

SECTION 710
DRAINAGE SYSTEM SIZING

710.1 Maximum fixture unit load. The maximum number of drainage fixture units connected to a given size of building sewer, building drain or horizontal branch of the building drain shall be determined using Table 710.1(1). The maximum number of drainage fixture units connected to a given size of horizontal branch or vertical soil or waste stack shall be determined using Table 710.1(2).

710.1.1 Horizontal stack offsets. Horizontal stack offsets shall be sized as required for building drains in accordance with Table 710.1(1), except as required by Section 711.4.

710.1.2 Vertical stack offsets. Vertical stack offsets shall be sized as required for straight stacks in accordance with Table 710.1(2), except where required to be sized as a building drain in accordance with Section 711.1.1.

710.2 Future fixtures. Where provision is made for the future installation of fixtures, those provided for shall be considered in determining the required sizes of drain pipes.

SECTION 711
OFFSETS IN DRAINAGE PIPING IN BUILDINGS OF FIVE STORIES OR MORE

711.1 Horizontal branch connections above or below vertical stack offsets. If a horizontal branch connects to the stack within 2 feet (610 mm) above or below a vertical stack offset, and the offset is located more than four branch intervals below the top of the stack, the offset shall be vented in accordance with Section 915.

711.1.1 Omission of vents for vertical stack offsets. Vents for vertical offsets required by Section 711.1 shall not be required where the stack and its offset are sized as a building drain [see Table 710.1(1), Column 5].

711.2 Horizontal branch connections to horizontal stack offsets. Where a horizontal stack offset is located more than four branch intervals below the top of the stack, a horizontal branch shall not connect within the horizontal stack offset or within 2 feet (610 mm) above or below such offset.

711.3 Horizontal stack offsets. A stack with a horizontal offset located more than four branch intervals below the top of the stack shall be vented in accordance with Section 915 and sized as follows:

1. The portion of the stack above the offset shall be sized as for a vertical stack based on the total number of drainage fixture units above the offset.

2. The offset shall be sized in accordance with Section 710.1.1.

3. The portion of the stack below the offset shall be sized as for the offset or based on the total number of drainage fixture units on the entire stack, whichever is larger [see Table 710.1(2), Column 4].

711.3.1 Omission of vents for horizontal stack offsets. Vents for horizontal stack offsets required by Section 711.3 shall not be required where the stack and its offset are one pipe size larger than required for a building drain [see Table 710.1(1), Column 5] and the entire stack and offset are not less in cross-sectional area than that required for a straight stack plus the area of an offset vent as provided for in Section 915. Omission of offset vents in accordance with this section shall not constitute approval of horizontal branch connections within the offset or within 2 feet (610 mm) above or below the offset.

711.4 Offsets below lowest branch. Where a vertical offset occurs in a soil or waste stack below the lowest horizontal branch, change in diameter of the stack because of the offset shall not be required. If a horizontal offset occurs in a soil or waste stack below the lowest horizontal branch, the required diameter of the offset and the stack below it shall be determined as for a building drain in accordance with Table 710.1(1).

SECTION 712
SUMPS AND EJECTORS

712.1 Building subdrains. Building subdrains that cannot be discharged to the sewer by gravity flow shall be discharged into a tightly covered and vented sump from which the liquid shall be lifted and discharged into the building gravity drainage system by automatic pumping equipment or other approved method. In other than existing structures, the sump shall not receive drainage from any piping within the building capable of being discharged by gravity to the building sewer.

712.2 Valves required. A check valve and full open valve, located on the discharge side of the check valve, shall be installed in the pump or ejector discharge piping between the pump or ejector and the gravity drainage system. Access shall be provided to such valves. Such valves will be located above the sump cover required by Section 712.1 or, where the discharge pipe from the ejector is below grade, the valves shall be accessibly located outside the sump below grade in an access pit with a removable access cover.

Exception: In one- and two-family dwellings, only a check valve shall be required, located on the discharge piping from the sewage pump or ejector.

712.3 Sump design. The sump pump, pit and discharge piping shall conform to the requirements of Sections 712.3.1 through 712.3.5.

712.3.1 Sump pump. The sump pump capacity and head shall be appropriate to anticipated use requirements.

TABLE 710.1 (1)
BUILDING DRAINS AND SEWERS

DIAMETER OF PIPE (inches)	MAXIMUM NUMBER OF DRAINAGE FIXTURE UNITS CONNECTED TO ANY PORTION OF THE BUILDING DRAIN OR THE BUILDING SEWER, INCLUDING BRANCHES OF THE BUILDING DRAIN[a]			
	Slope per foot			
	$^1/_{16}$ inch	$^1/_8$ inch	¼ inch	½ inch
$1^1/_4$	—	—	1	1
$1^1/_2$	—	—	3	3
2	—	—	21	26
$2^1/_2$	—	—	24	31
3	—	36	42	50
4	—	180	216	250
5	—	390	480	575
6	—	700	840	1,000
8	1,400	1,600	1,920	2,300
10	2,500	2,900	3,500	4,200
12	3,900	4,600	5,600	6,700
15	7,000	8,300	10,000	12,000

For SI: 1 inch = 25.4 mm, 1 inch per foot = 83.3 mm/m.

a. The minimum size of any building drain serving a water closet shall be 3 inches.

TABLE 710.1(2)
HORIZONTAL FIXTURE BRANCHES AND STACKS [a]

DIAMETER OF PIPE (inches)	MAXIMUM NUMBER OF DRAINAGE FIXTURE UNITS (dfu)			
	Total for horizotal branch	Stacks[b]		
		Total discharge into one branch interval	Total for stack of three branch Intervals or less	Total for stack Greater than three Branch intervals
$1^1/_2$	3	2	4	8
2	6	6	10	24
$2^1/_2$	12	9	20	42
3	20	20	48	72
4	160	90	240	500
5	360	200	540	1,100
6	620	350	960	1,900
8	1,400	600	2,200	3,600
10	2,500	1,000	3,800	5,600
12	2,900	1,500	6,000	8,400
15	7,000	Note c	Note c	Note c

For SI:1 inch = 25.4mm

a. Does not include branches of the building drain. Refer to Table 710.1(1).

b. Stacks shall be sized based on the total accumulated connected load at each story or branch interval. As the total accumulated connected load decreases, stacks are permitted to be reduced in size. Stack diameters shall not be reduced to less than one-half of the diameter of the largest stack size required.

c. Sizing load based on design criteria.

712.3.2 Sump pit. The sump pit shall be not less than 18 inches (457 mm) in diameter and 24 inches (610 mm) deep, unless otherwise approved. The pit shall be accessible and located such that all drainage flows into the pit by gravity. The sump pit shall be constructed of tile, concrete, steel, plastic or other approved materials. The pit bottom shall be solid and provide permanent support for the pump. The sump pit shall be fitted with a gas-tight removable cover adequate to support anticipated loads in the area of use. The sump pit shall be vented in accordance with Chapter 9.

712.3.3 Discharge piping. Discharge piping shall meet the requirements of Section 712.2.

712.3.4 Maximum effluent level. The effluent level control shall be adjusted and maintained to at all times prevent the effluent in the sump from rising to within 2 inches (51 mm) of the invert of the gravity drain inlet into the sump.

712.3.5 Ejector connection to the drainage system. Pumps connected to the drainage system shall connect to the building sewer or shall connect to a wye fitting in the building drain a minimum of 10 feet (3048 mm) from the base of any soil stack, waste stack or fixture drain. Where the discharge line connects into horizontal drainage piping, the connector shall be made through a wye fitting into the top of the drainage piping.

712.4 Sewage pumps and sewage ejectors. A sewage pump or sewage ejector shall automatically discharge the contents of the sump to the building drainage system.

712.4.1 Macerating toilet systems. Macerating toilet systems shall comply with CSA B45.9 or ASME A112.3.4 and shall be installed in accordance with the manufacturer's installation instructions.

712.4.2 Capacity. A sewage pump or sewage ejector shall have the capacity and head for the application requirements. Pumps or ejectors that receive the discharge of water closets shall be capable of handling spherical solids with a diameter of up to and including 2 inches (51 mm). Other pumps or ejectors shall be capable of handling spherical solids with a diameter of up to and including 1 inch (25.4 mm). The minimum capacity of a pump or ejector based on the diameter of the discharge pipe shall be in accordance with Table 712.4.2.

Exceptions:

1. Grinder pumps or grinder ejectors that receive the discharge of water closets shall have a minimum discharge opening of 1.25 inches (32 mm).

2. Macerating toilet assemblies that serve single water closets shall have a minimum discharge opening of 0.75 inch (19 mm).

SECTION 713
HEALTH CARE PLUMBING

713.1 Scope. This section shall govern those aspects of health care plumbing systems that differ from plumbing systems in other structures. Health care plumbing systems shall conform to this section in addition to the other requirements of this code. The provisions of this section shall apply to the special devices and equipment installed and maintained in the following occupancies: nursing homes; homes for the aged; orphanages; infirmaries; first aid stations; psychiatric facilities; clinics; professional offices of dentists and doctors; mortuaries; educational facilities; surgery, dentistry, research and testing laboratories; establishments manufacturing pharmaceutical drugs and medicines; and other structures with similar apparatus and equipment classified as plumbing.

TABLE 712.4.2
MINIMUM CAPACITY OF SEWAGE PUMP
OR SEWAGE EJECTOR

DIAMETER OF THE DISCHARGE PIPE (inches)	CAPACITY OF PUMP OR EJECTOR (gpm)
2	21
2½	30
3	46

For SI:1 inch = 25.4 mm, 1 gallon per minute = 3.785 L/m.

713.2 Bedpan washers and clinical sinks. Bedpan washers and clinical sinks shall connect to the drainage and vent system in accordance with the requirements for a water closet. Bedpan washers shall also connect to a local vent.

713.3 Indirect waste. All sterilizers, steamers and condensers shall discharge to the drainage through an indirect waste pipe by means of an air gap. Where a battery of not more than three sterilizers discharges to an individual receptor, the distance between the receptor and a sterilizer shall not exceed 8 feet (2438 mm). The indirect waste pipe on a bedpan steamer shall be trapped.

713.4 Vacuum system station. Ready access shall be provided to vacuum system station receptacles. Such receptacles shall be built into cabinets or recesses and shall be visible.

713.5 Bottle system. Vacuum (fluid suction) systems intended for collecting, removing and disposing of blood, pus or other fluids by the bottle system shall be provided with receptacles equipped with an overflow prevention device at each vacuum outlet station.

713.6 Central disposal system equipment. All central vacuum (fluid suction) systems shall provide continuous service. Systems equipped with collecting or control tanks shall provide for draining and cleaning of the tanks while the system is in operation. In hospitals, the system shall be connected to the emergency power system. The exhausts from a vacuum pump serving a vacuum (fluid suction) system shall discharge separately to open air above the roof.

713.7 Central vacuum or disposal systems. Where the waste from a central vacuum (fluid suction) system of the barometric-lag, collection-tank or bottle-disposal type is connected to the drainage system, the waste shall be directly connected to the sanitary drainage system through a trapped waste.

713.7.1 Piping. The piping of a central vacuum (fluid suction) system shall be of corrosion-resistant material with a smooth interior surface. A branch shall not be less than 0.5-inch (12.7 mm) nominal pipe size for one outlet and shall be sized in accordance with the number of vacuum outlets. A main shall not be less than 1-inch (25 mm) nominal pipe size. The pipe sizing shall be increased in accordance with the manufacturer's instructions as stations are increased.

713.7.2 Velocity. The velocity of airflow in a central vacuum (fluid suction) system shall be less than 5,000 feet per minute (25 m/s).

713.8 Vent connections prohibited. Connections between local vents serving bedpan washers or sterilizer vents serving sterilizing apparatus and normal sanitary plumbing systems are prohibited. Only one type of apparatus shall be served by a local vent.

713.9 Local vents and stacks for bedpan washers. Bedpan washers shall be vented to open air above the roof by means of one or more local vents. The local vent for a bedpan washer shall not be less than a 2-inch-diameter (51 mm) pipe. A local vent serving a single bedpan washer is permitted to drain to the fixture served.

713.9.1 Multiple installations. Where bedpan washers are located above each other on more than one floor, a local vent stack is permitted to be installed to receive the local vent on the various floors. Not more than three bedpan washers shall be connected to a 2-inch (51 mm) local vent stack, not more than six to a 3-inch (76 mm) local vent stack and not more than 12 to a 4-inch (102 mm) local vent stack. In multiple installations, the connections between a bedpan washer local vent and a local vent stack shall be made with tee or tee-wye sanitary pattern drainage fittings installed in an upright position.

713.9.2 Trap required. The bottom of the local vent stack, except where serving only one bedpan washer, shall be drained by means of a trapped and vented waste connection to the sanitary drainage system. The trap and waste shall be the same size as the local vent stack.

713.9.3 Trap seal maintenance. A water supply pipe not less than $\frac{1}{4}$ inch (6.4 mm) in diameter shall be taken from the flush supply of each bedpan washer on the discharge or fixture side of the vacuum breaker, shall be trapped to form not less than a 3-inch (76 mm) water seal, and shall be connected to the local vent stack on each floor. The water supply shall be installed so as to provide a supply of water to the local vent stack for cleansing and drain trap seal maintenance each time a bedpan washer is flushed.

713.10 Sterilizer vents and stacks. Multiple installations of pressure and nonpressure sterilizers shall have the vent connections to the sterilizer vent stack made by means of inverted wye fittings. Access shall be provided to vent connections for the purpose of inspection and maintenance.

713.10.1 Drainage. The connection between sterilizer vent or exhaust openings and the sterilizer vent stack shall be designed and installed to drain to the funnel or basket-type waste fitting. In multiple installations, the sterilizer vent stack shall be drained separately to the lowest sterilizer funnel or basket-type waste fitting or receptor.

713.11 Sterilizer vent stack sizes. Sterilizer vent stack sizes shall comply with Sections 713.11.1 through 713.11.4.

713.11.1 Bedpan steamers. The minimum size of a sterilizer vent serving a bedpan steamer shall be 1.50 inches (38

mm) in diameter. Multiple installations shall be sized in accordance with Table 713.11.1.

TABLE 713.11.1
STACK SIZES FOR BEDPAN STEAMERS AND BOILING-TYPE STERILIZERS
(Number of Connections of Various Sizes Permitted to Various-sized Sterilizer Vent Stacks)

STACK SIZE (inches)	CONNECTION SIZE		
	$1^1/_2''$		2''
$1^1/_2{}^a$	1	or	0
2^a	2	or	1
2^b	1	and	1
3^a	4	or	2
3^b	2	and	2
4^a	8	or	4
4^b	4	and	4

For SI: 1 inch = 25.4 mm.

a. Total of each size.

b. Combination of sizes.

713.11.2 Boiling-type sterilizers. The minimum size of a sterilizer vent stack shall be 2 inches (51 mm) in diameter where serving a utensil sterilizer and 1.5 inches (38 mm) in diameter where serving an instrument sterilizer. Combinations of boiling-type sterilizer vent connections shall be sized in accordance with Table 713.11.1.

713.11.3 Pressure sterilizers. Pressure sterilizer vent stacks shall be 2.5 inches (64 mm) minimum. Those serving combinations of pressure sterilizer exhaust connections shall be sized in accordance with Table 713.11.3.

713.11.4 Pressure instrument washer sterilizer sizes. The minimum diameter of a sterilizer vent stack serving an instrument washer sterilizer shall be 2 inches (51 mm). Not more than two sterilizers shall be installed on a 2-inch (51 mm) stack, and not more than four sterilizers shall be installed on a 3-inch (76 mm) stack.

SECTION 714
COMPUTERIZED DRAINAGE DESIGN

714.1 Design of drainage system. The sizing, design and layout of the drainage system shall be permitted to be designed by approved computer design methods.

714.2 Load on drainage system. The load shall be computed from the simultaneous or sequential discharge conditions from fixtures, appurtenances and appliances or the peak usage design condition.

714.2.1 Fixture discharge profiles. The discharge profiles for flow rates versus time from fixtures and appliances shall be in accordance with the manufacturer's specifications.

714.3 Selections of drainage pipe sizes. Pipe shall be sized to prevent full-bore flow.

714.3.1 Selecting pipe wall roughness. Pipe size calculations shall be conducted with the pipe wall roughness factor

(k_s), in accordance with the manufacturer's specifications and as modified for aging roughness factors with deposits and corrosion.

TABLE 713.11.3
STACK SIZES FOR PRESSURE STERILIZERS
(Number of Connections of Various Sizes Permitted
To Various-sized Vent Stacks)

STACK SIZE (inches)	CONNECTION SIZE			
	$^3/_4''$	$1''$	$1^1/_4''$	$1^1/_2''$
$1^1/_2{}^a$	3 or	2 or	1	—
$1^1/_2{}^b$	2 and	1	—	—
2^a	6 or	3 or	2 or	1
2^b	3 and	2	—	—
2^b	2 and	1 and	1	—
2^b	1 and	1 and	—	1
3^a	15 or	7 or	5 or	3
3^b	1 and	1 and 5 and	2 and —	2 1

For SI: 1 inch = 25.4 mm.

a. Total of each size.

b. Combination of sizes.

714.3.2 Slope of horizontal drainage piping. Horizontal drainage piping shall be designed and installed at slopes in accordance with Table 704.1.

SECTION 715
BACKWATER VALVES

715.1 Sewage backflow. Where the flood level rims of plumbing fixtures are below the elevation of the manhole cover of the next upstream manhole in the public sewer, such fixtures shall be protected by a backwater valve installed in the building drain, branch of the building drain or horizontal branch serving such fixtures. Plumbing fixtures having flood level rims above the elevation of the manhole cover of the next upstream manhole in the public sewer shall not discharge through a backwater valve.

715.2 Material. All bearing parts of backwater valves shall be of corrosion-resistant material. Backwater valves shall comply with ASME A112.14.1, CSA B181.1 or CSA B181.2.

715.3 Seal. Backwater valves shall be so constructed as to provide a mechanical seal against backflow.

715.4 Diameter. Backwater valves, when fully opened, shall have a capacity not less than that of the pipes in which they are installed.

715.5 Location. Backwater valves shall be installed so that access is provided to the working parts for service and repair.

CHAPTER 8
INDIRECT/SPECIAL WASTE

SECTION 801
GENERAL

801.1 Scope. This chapter shall govern matters concerning indirect waste piping and special wastes. This chapter shall further control matters concerning food-handling establishments, sterilizers, clear-water wastes, swimming pools, methods of providing air breaks or air gaps, and neutralizing devices for corrosive wastes.

801.2 Protection. All devices, appurtenances, appliances and apparatus intended to serve some special function, such as sterilization, distillation, processing, cooling, or storage of ice or foods, and that discharge to the drainage system, shall be provided with protection against backflow, flooding, fouling, contamination and stoppage of the drain.

SECTION 802
INDIRECT WASTES

802.1 Where required. Food-handling equipment and clear-water waste shall discharge through an indirect waste pipe as specified in Sections 802.1.1 through 802.1.7. All health-care related fixtures, devices and equipment shall discharge to the drainage system through an indirect waste pipe by means of an air gap in accordance with this chapter and Section 713.3. Fixtures not required by this section to be indirectly connected shall be directly connected to the plumbing system in accordance with Chapter 7.

802.1.1 Food handling. Equipment and fixtures utilized for the storage, preparation and handling of food shall discharge through an indirect waste pipe by means of an air gap.

802.1.2 Floor drains in food storage areas. Floor drains located within walk-in refrigerators or freezers in food service and food establishments shall be indirectly connected to the sanitary drainage system by means of an air gap. Where a floor drain is located within an area subject to freezing, the waste line serving the floor drain shall not be trapped and shall indirectly discharge into a waste receptor located outside of the area subject to freezing.

> **Exception:** Where protected against backflow by a backwater valve, such floor drains shall be indirectly connected to the sanitary drainage system by means of an air break or an air gap.

802.1.3 Potable clear-water waste. Where devices and equipment, such as sterilizers and relief valves, discharge potable water to the building drainage system, the discharge shall be through an indirect waste pipe by means of an air gap.

802.1.4 Swimming pools. Where wastewater from swimming pools, backwash from filters and water from pool deck drains discharge to the building drainage system, the discharge shall be through an indirect waste pipe by means of an air gap.

802.1.5 Nonpotable clear-water waste. Where devices and equipment such as process tanks, filters, drips and boilers discharge nonpotable water to the building drainage system, the discharge shall be through an indirect waste pipe by means of an air break or an air gap.

802.1.6 Domestic Dishwashing machines. Domestic dishwashing machines shall discharge indirectly through an air gap or air break into a standpipe or waste receptor in accordance with Section 802.2, or discharge into a wye-branch fitting on the tailpiece of the kitchen sink or the dishwasher connection of a food waste grinder. The waste line of a domestic dishwashing machine discharging into a kitchen sink tailpiece or food waste grinder shall connect to a deck-mounted air gap or the waste line shall rise and be securely fastenend to the underside of the sink rim or counter.

802.1.7 Commercial dishwashing machines. The discharge from a commercial dishwashing machine shall be through an air gap or air break into a standpipe or waste receptor in accordance with Section 802.2.

802.2 Installation. All indirect waste piping shall discharge through an air gap or air break into a waste receptor or standpipe. Waste receptors and standpipes shall be trapped and vented and shall connect to the building drainage system. All indirect waste piping that exceeds 2 feet (610 mm) in developed length measured horizontally, or 4 feet (1219 mm) in total developed length, shall be trapped.

802.2.1 Air gap. The air gap between the indirect waste pipe and the flood level rim of the waste receptor shall be a minimum of twice the effective opening of the indirect waste pipe.

802.2.2 Air break. An air break shall be provided between the indirect waste pipe and the trap seal of the waste receptor or standpipe.

802.3 Waste receptors. Every waste receptor shall be of an approved type. A removable strainer or basket shall cover the waste outlet of waste receptors. Waste receptors shall be installed in ventilated spaces. Waste receptors shall not be installed in bathrooms or toilet rooms or in any inaccessible or unventilated space such as a closet or storeroom. Ready access shall be provided to waste receptors.

802.3.1 Size of receptors. A waste receptor shall be sized for the maximum discharge of all indirect waste pipes served by the receptor. Receptors shall be installed to prevent splashing or flooding.

802.3.2 Open hub waste receptors. Waste receptors shall be permitted in the form of a hub or pipe extending not less than 1 inch (25.4 mm) above a water-impervious floor and are not required to have a strainer.

802.4 Standpipes. Standpipes shall be individually trapped. Standpipes shall extend a minimum of 18 inches (457 mm) and a maximum of 42 inches (1066 mm) above the trap weir. Access shall be provided to all standpipes and drains for rodding.

SECTION 803
SPECIAL WASTES

803.1 Wastewater temperature. Steam pipes shall not connect to any part of a drainage or plumbing system and water above 140°F (60°C) shall not be discharged into any part of a drainage system. Such pipes shall discharge into an indirect waste receptor connected to the drainage system.

803.2 Neutralizing device required for corrosive wastes. Corrosive liquids, spent acids or other harmful chemicals that destroy or injure a drain, sewer, soil or waste pipe, or create noxious or toxic fumes or interfere with sewage treatment processes shall not be discharged into the plumbing system without being thoroughly diluted, neutralized or treated by passing through an approved dilution or neutralizing device. Such devices shall be automatically provided with a sufficient supply of diluting water or neutralizing medium so as to make the contents noninjurious before discharge into the drainage system. The nature of the corrosive or harmful waste and the method of its treatment or dilution shall be approved prior to installation.

803.3 System design. A chemical drainage and vent system shall be designed and installed in accordance with this code. Chemical drainage and vent systems shall be completely separated from the sanitary systems. Chemical waste shall not discharge to a sanitary drainage system until such waste has been treated in accordance with Section 803.2.

SECTION 804
MATERIALS, JOINTS AND CONNECTIONS

804.1 General. The materials and methods utilized for the construction and installation of indirect waste pipes and systems shall comply with the applicable provisions of Chapter 7.

CHAPTER 9

VENTS

SECTION 901
GENERAL

901.1 Scope. The provisions of this chapter shall govern the materials, design, construction and installation of vent systems.

901.2 Trap seal protection. The plumbing system shall be provided with a system of vent piping that will permit the admission or emission of air so that the seal of any fixture trap shall not be subjected to a pneumatic pressure differential of more than 1 inch of water column (249 Pa).

901.2.1 Venting required. Every trap and trapped fixture shall be vented in accordance with one of the venting methods specified in this chapter.

901.3 Chemical waste vent system. The vent system for a chemical waste system shall be independent of the sanitary vent system and shall terminate separately through the roof to the open air.

901.4 Use limitations. The plumbing vent system shall not be utilized for purposes other than the venting of the plumbing system.

901.5 Tests. The vent system shall be tested in accordance with Section 312.

901.6 Engineered systems. Engineered venting systems shall conform to the provisions of Section 918.

SECTION 902
MATERIALS

902.1 Vents. The materials and methods utilized for the construction and installation of venting systems shall comply with the applicable provisions of Section 702.

902.2 Sheet copper. Sheet copper for vent pipe flashings shall conform to ASTM B 152 and shall weigh not less than 8 ounces per square foot (2.5 kg/m²).

902.3 Sheet lead. Sheet lead for vent pipe flashings shall weigh not less than 3 pounds per square foot (15 kg/m²) for field-constructed flashings and not less than 2.5 pounds per square foot (12 kg/m²) for prefabricated flashings.

SECTION 903
VENT STACKS AND STACK VENTS

903.1 Stack required. Every building in which plumbing is installed shall have at least one stack the size of which is not less than one-half of the required size of the building drain. Such stack shall run undiminished in size and as directly as possible from the building drain through to the open air or to a vent header that extends to the open air.

903.1.1 Connection to drainage system. A vent stack shall connect to the building drain or to the base of a drainage stack in accordance with Section 903.4. A stack vent shall be an extension of the drainage stack.

903.2 Vent stack required. A vent stack shall be required for every drainage stack that is five branch intervals or more.

903.3 Vent termination. Every vent stack or stack vent shall extend outdoors and terminate to the open air.

903.4 Vent connection at base. Every vent stack shall connect to the base of the drainage stack. The vent stack shall connect at or below the lowest horizontal branch. Where the vent stack connects to the building drain, the connection shall be located downstream of the drainage stack and within a distance of 10 times the diameter of the drainage stack.

903.5 Vent headers. Stack vents and vent stacks connected into a common vent header at the top of the stacks and extending to the open air at one point shall be sized in accordance with the requirements of Section 916.1. The number of fixture units shall be the sum of all fixture units on all stacks connected thereto, and the developed length shall be the longest vent length from the intersection at the base of the most distant stack to the vent terminal in the open air, as a direct extension of one stack.

SECTION 904
VENT TERMINALS

904.1 Roof extension. All open vent pipes that extend through a roof shall be terminated at least [NUMBER] inches (mm) above the roof, except that where a roof is to be used for any purpose other than weather protection, the vent extensions shall be run at least 7 feet (2134 mm) above the roof.

904.2 Frost closure. Where the 97.5-percent value for outside design temperature is 0°F (-18°C) or less, every vent extension through a roof or wall shall be a minimum of 3 inches (76 mm) in diameter. Any increase in the size of the vent shall be made inside the structure a minimum of 1 foot (305 mm) below the roof or inside the wall.

904.3 Flashings. The juncture of each vent pipe with the roof line shall be made water tight by an approved flashing.

904.4 Prohibited use. Vent terminals shall not be used as a flag pole or to support flag poles, television aerials or similar items, except when the piping has been anchored in an approved manner.

904.5 Location of vent terminal. An open vent terminal from a drainage system shall not be located directly beneath any door, openable window, or other air intake opening of the building or of an adjacent building, and any such vent terminal shall not be within 10 feet (3048 mm) horizontally of such an opening unless it is at least 2 feet (610 mm) above the top of such opening.

904.6 Extension through the wall. Vent terminals extending through the wall shall terminate a minimum of 10 feet (3048 mm) from the lot line and 10 feet (3048 mm) above average ground level. Vent terminals shall not terminate under the overhang of a structure with soffit vents. Side wall vent terminals shall be protected to prevent birds or rodents from entering or blocking the vent opening.

904.7 Extension outside a structure. In climates where the 97.5-percent value for outside design temperature is less than 0°F (-18°C), vent pipes installed on the exterior of the structure shall be protected against freezing by insulation, heat or both.

SECTION 905
VENT CONNECTIONS AND GRADES

905.1 Connection. All individual, branch and circuit vents shall connect to a vent stack, stack vent, air admittance valve or extend to the open air.

905.2 Grade. All vent and branch vent pipes shall be so graded and connected as to drain back to the drainage pipe by gravity.

905.3 Vent connection to drainage system. Every dry vent connecting to a horizontal drain shall connect above the centerline of the horizontal drain pipe.

905.4 Vertical rise of vent. Every dry vent shall rise vertically to a minimum of 6 inches (152 mm) above the flood level rim of the highest trap or trapped fixture being vented.

Exception: Vents for interceptors located outdoors.

905.5 Height above fixtures. A connection between a vent pipe and a vent stack or stack vent shall be made at least 6 inches (152 mm) above the flood level rim of the highest fixture served by the vent. Horizontal vent pipes forming branch vents, relief vents or loop vents shall be at least 6 inches (152 mm) above the flood level rim of the highest fixture served.

905.6 Vent for future fixtures. Where the drainage piping has been roughed-in for future fixtures, a rough-in connection for a vent shall be installed. The vent size shall be not less than one-half the diameter of the rough-in drain to be served. The vent rough-in shall connect to the vent system, or shall be vented by other means as provided for in this chapter. The connection shall be identified to indicate that it is a vent.

SECTION 906
FIXTURE VENTS

906.1 Distance of trap from vent. Each fixture trap shall have a protecting vent located so that the slope and the developed length in the fixture drain from the trap weir to the vent fitting are within the requirements set forth in Table 906.1.

906.2 Venting of fixture drains. The vent for a fixture drain, except where serving a fixture with integral traps, such as water closets, shall connect above the weir of the fixture trap being vented.

906.3 Crown vent. A vent shall not be installed within two pipe diameters of the trap weir.

SECTION 907
INDIVIDUAL VENT

907.1 Individual vent permitted. Each trap and trapped fixture is permitted to be provided with an individual vent. The individual vent shall connect to the fixture drain of the trap or trapped fixture being vented.

TABLE 906.1
MAXIMUM DISTANCE OF FIXTURE TRAP FROM VENT

SIZE OF TRAP (inches)	SIZE OF FIXTURE DRAIN (inches)	SLOPE (inch per foot)	DISTANCE FROM TRAP (feet)
$1^1/_4$	$1^1/_4$	$^1/_4$	$3^1/_2$
$1^1/_4$	$1^1/_2$	$^1/_4$	5
$1^1/_2$	$1^1/_2$	$^1/_4$	5
$1^1/_2$	2	$^1/_4$	6
2	2	$^1/_4$	6
3	3	$^1/_8$	10
4	4	$^1/_8$	12

For SI: 1 inch = 25.4 mm, 1 foot = 304.8 mm, 1 inch per foot = 83.3 mm/m.

SECTION 908
COMMON VENT

908.1 Individual vent as common vent. An individual vent is permitted to vent two traps or trapped fixtures as a common vent. The traps or trapped fixtures being common vented shall be located on the same floor level.

908.2 Connection at the same level. Where the fixture drains being common vented connect at the same level, the vent connection shall be at the interconnection of the fixture drains or downstream of the interconnection.

908.3 Connection at different levels. Where the fixture drains connect at different levels, the vent shall connect as a vertical extension of the vertical drain. The vertical drain pipe connecting the two fixture drains shall be considered the vent for the lower fixture drain, and shall be sized in accordance with Table 908.3. The upper fixture shall not be a water closet.

TABLE 908.3
COMMON VENT SIZES

PIPE SIZE (inches)	MAXIMUM DISCHARGE FROM UPPER FIXTURE DRAIN (dfu)
$1^1/_2$	1
2	4
$2^1/_2$ to 3	6

For SI: 1 inch = 25.4 mm.

SECTION 909
WET VENTING

909.1 Wet vent permitted. Any combination of fixtures within two bathroom groups located on the same floor level are permitted to be vented by a wet vent. The wet vent shall be considered the vent for the fixtures and shall extend from the connection of the dry vent along the direction of the flow in the drain pipe to the most downstream fixture drain connection to the horizontal branch drain. Only the fixtures within the bathroom groups shall connect to the wet-vented horizontal branch drain. Any additional fixtures shall discharge downstream of the wet vent.

909.1.1 Vertical wet vent. Any combination of fixtures within two bathroom groups located on the same floor level is permitted to be vented by a vertical wet vent. The vertical wet vent shall extend from the connection to the dry vent down to the lowest fixture drain connection. Each fixture shall connect independently to the vertical wet vent. Water closet drains shall connect at the same elevation. Other fixture drains shall connect above or at the same elevation as the water closet fixture drains. The dry vent connection to the vertical wet vent shall be an individual or common vent serving one or two fixtures.

909.2 Vent connection. The dry vent connection to the wet vent shall be an individual vent or common vent to the lavatory, bidet, shower or bathtub. The dry vent shall be sized based on the largest required diameter of pipe within the wet vent system served by the dry vent.

909.3 Size. The wet vent shall be of a minimum size as specified in Table 909.3, based on the fixture unit discharge to the wet vent.

TABLE 909.3
WET VENT SIZE

WET VENT PIPE SIZE (inches)	DRAINAGE FIXTURE UNIT LOAD (dfu)
$1^1/_2$	1
2	4
$2^1/_2$	6
3	12

For SI: 1 inch = 25.4 mm.

SECTION 910
WASTE STACK VENT

910.1 Waste stack vent permitted. A waste stack shall be considered a vent for all of the fixtures discharging to the stack where installed in accordance with the requirements of this section.

910.2 Stack installation. The waste stack shall be vertical, and both horizontal and vertical offsets shall be prohibited. Every fixture drain shall connect separately to the waste stack. The stack shall not receive the discharge of water closets or urinals.

910.3 Stack vent. A stack vent shall be provided for the waste stack. The size of the stack vent shall be equal to the size of the waste stack. Offsets shall be permitted in the stack vent and shall be located at least 6 inches (152 mm) above the flood level of the highest fixture, and shall be in accordance with Section 905.2.

910.4 Waste stack size. The waste stack shall be sized based on the total discharge to the stack and the discharge within a branch interval in accordance with Table 910.4. The waste stack shall be the same size throughout its length.

TABLE 910.4
WASTE STACK VENT SIZE

STACK SIZE (inches)	MAXIMUM NUMBER OF DRAINAGE FIXTURE UNITS (dfu)	
	Total discharge into one branch interval	Total discharge for stack
$1^1/_2$	1	2
2	2	4
$2^1/_2$	No limit	8
3	No limit	24
4	No limit	50
5	No limit	75
6	No limit	100

For SI: 1 inch = 25.4 mm.

SECTION 911
CIRCUIT VENTING

911.1 Circuit vent permitted. A maximum of eight fixtures connected to a horizontal branch drain shall be permitted to be circuit vented. Each fixture drain shall connect horizontally to the horizontal branch being circuit vented. The horizontal branch drain shall be classified as a vent from the most downstream fixture drain connection to the most upstream fixture drain connection to the horizontal branch.

911.1.1 Multiple circuit-vented branches. Circuit-vented horizontal branch drains are permitted to be connected together. Each group of a maximum of eight fixtures shall be considered a separate circuit vent and shall conform to the requirements of this section.

911.2 Vent connection. The circuit vent connection shall be located between the two most upstream fixture drains. The vent shall connect to the horizontal branch and shall be installed in accordance with Section 905. The circuit vent pipe shall not receive the discharge of any soil or waste.

911.3 Slope and size of horizontal branch. The maximum slope of the vent section of the horizontal branch drain shall be one unit vertical in 12 units horizontal (8-percent slope). The entire length of the vent section of the horizontal branch drain shall be sized for the total drainage discharge to the branch.

911.3.1 Size of multiple circuit vent. Each separate circuit-vented horizontal branch that is interconnected shall be sized independently in accordance with Section 911.3. The downstream circuit-vented horizontal branch shall be sized for the total discharge into the branch, including the upstream branches and the fixtures within the branch.

911.4 Relief vent. A relief vent shall be provided for circuit-vented horizontal branches receiving the discharge of four or more water closets and connecting to a drainage stack that receives the discharge of soil or waste from upper horizontal branches.

911.4.1 Connection and installation. The relief vent shall connect to the horizontal branch drain between the stack and the most downstream fixture drain of the circuit vent. The relief vent shall be installed in accordance with Section 905.

911.4.2 Fixture drain or branch. The relief vent is permitted to be a fixture drain or fixture branch for fixtures located within the same branch interval as the circuit-vented horizontal branch. The maximum discharge to a relief vent shall be four fixture units.

911.5 Additional fixtures. Fixtures, other than the circuit-vented fixtures, are permitted to discharge to the horizontal branch drain. Such fixtures shall be located on the same floor as the circuit-vented fixtures and shall be either individually or common vented.

SECTION 912
COMBINATION DRAIN AND VENT SYSTEM

912.1 Type of fixtures. A combination drain and vent system shall not serve fixtures other than floor drains, sinks, lavatories and drinking fountains. Combination drain and vent systems shall not receive the discharge from a food waste grinder or clinical sink.

912.2 Installation. The only vertical pipe of a combination drain and vent system shall be the connection between the fixture drain of a sink, lavatory or drinking fountain, and the horizontal combination drain and vent pipe. The maximum vertical distance shall be 8 feet (2438 mm).

912.2.1 Slope. The horizontal combination drain and vent pipe shall have a maximum slope of one-half unit vertical in 12 units horizontal (4-percent slope). The minimum slope shall be in accordance with Table 704.1.

912.2.2 Connection. The combination drain and vent system shall be provided with a dry vent connected at any point within the system or the system shall connect to a horizontal drain that is vented in accordance with one of the venting methods specified in this chapter. Combination drain and vent systems connecting to building drains receiving only the discharge from a stack or stacks shall be provided with a dry vent. The vent connection to the combination drain and vent pipe shall extend vertically a minimum of 6 inches (152 mm) above the flood level rim of the highest fixture being vented before offsetting horizontally.

912.2.3 Vent size. The vent shall be sized for the total drainage fixture unit load in accordance with Section 916.2.

912.2.4 Fixture branch or drain. The fixture branch or fixture drain shall connect to the combination drain and vent within a distance specified in Table 906.1. The combination drain and vent pipe shall be considered the vent for the fixture.

912.3 Size. The minimum size of a combination drain and vent pipe shall be in accordance with Table 912.3.

TABLE 912.3
SIZE OF COMBINATION DRAIN AND VENT PIPE

DIAMETER PIPE (inches)	MAXIMUM NUMBER OF DRAINAGE FIXTURE UNITS (dfu)	
	Connecting to a horizontal branch or stack	Connecting to a building drain or building subdrain
2	3	4
$2^1/_2$	6	26
3	12	31
4	20	50
5	160	250
6	360	575

SI: 1 inch = 25.4 mm.

SECTION 913
ISLAND FIXTURE VENTING

913.1 Limitation. Island fixture venting shall not be permitted for fixtures other than sinks and lavatories. Residential kitchen sinks with a dishwasher waste connection, a food waste grinder, or both, in combination with the kitchen sink waste, shall be permitted to be vented in accordance with this section.

913.2 Vent connection. The island fixture vent shall connect to the fixture drain as required for an individual or common vent. The vent shall rise vertically to above the drainage outlet of the fixture being vented before offsetting horizontally or vertically downward. The vent or branch vent for multiple island fixture vents shall extend to a minimum of 6 inches (152 mm) above the highest island fixture being vented before connecting to the outside vent terminal.

913.3 Vent installation below the fixture flood level rim. The vent located below the flood level rim of the fixture being vented shall be installed as required for drainage piping in accordance with Chapter 7, except for sizing. The vent shall be sized in accordance with Section 916.2. The lowest point of the island fixture vent shall connect full size to the drainage system. The connection shall be to a vertical drain pipe or to the top half of a horizontal drain pipe. Cleanouts shall be provided in the island fixture vent to permit rodding of all vent piping located below the flood level rim of the fixtures. Rodding in both directions shall be permitted through a cleanout.

SECTION 914
RELIEF VENTS—STACKS OF MORE THAN
10 BRANCH INTERVALS

914.1 Where required. Soil and waste stacks in buildings having more than 10 branch intervals shall be provided with a relief vent at each tenth interval installed, beginning with the top floor.

914.2 Size and connection. The size of the relief vent shall be equal to the size of the vent stack to which it connects. The lower end of each relief vent shall connect to the soil or waste stack through a wye below the horizontal branch serving the floor, and the upper end shall connect to the vent stack through a wye not less than 3 feet (914 mm) above the floor.

SECTION 915
VENTS FOR STACK OFFSETS

915.1 Vent for horizontal offset of drainage stack. Horizontal offsets of drainage stacks shall be vented where five or more branch intervals are located above the offset. The offset shall be vented by venting the upper section of the drainage stack and the lower section of the drainage stack.

915.2 Upper section. The upper section of the drainage stack shall be vented as a separate stack with a vent stack connection

installed in accordance with Section 903.4. The offset shall be considered the base of the stack.

915.3 Lower section. The lower section of the drainage stack shall be vented by a yoke vent connecting between the offset and the next lower horizontal branch. The yoke vent connection shall be permitted to be a vertical extension of the drainage stack. The size of the yoke vent and connection shall be a minimum of the size required for the vent stack of the drainage stack.

SECTION 916
VENT PIPE SIZING

916.1 Size of stack vents and vent stacks. The minimum required diameter of stack vents and vent stacks shall be determined from the developed length and the total of drainage fixture units connected thereto in accordance with Table 916.1, but in no case shall the diameter be less than one-half the diameter of the drain served or less than $1^{1}/_{4}$ inches (32 mm).

TABLE 916.1
SIZE AND DEVELOPED LENGTH OF STACK VENTS AND VENT STACKS

DIAMETER OF SOIL OR WASTE STACK (inches)	TOTAL FIXTURE UNITS BEING VENTED (dfu)	MAXIMUM DEVELOPED LENGTH OF VENT (feet)[a] DIAMETER OF VENT (inches)										
		$1^{1}/_{4}$	$1^{1}/_{2}$	2	$2^{1}/_{2}$	3	4	5	6	8	10	12
$1^{1}/_{4}$	2	30										
$1^{1}/_{2}$	8	50	150	—	—	—	—	—	—	—	—	—
$1^{1}/_{2}$	10	30	100									
2	12	30	75	200								
2	20	26	50	150	—	—	—	—	—	—	—	—
$2^{1}/_{2}$	42		30	100	300							
3	10		42	150	360	1,040						
3	21	—	32	110	270	810	—	—	—	—	—	—
3	53		27	94	230	680						
3	102		25	86	210	620						
4	43	—		35	85	250	980	—	—	—	—	—
4	140			27	65	200	750					
4	320			23	55	170	640					
4	540	—	—	21	50	150	580					
5	190				28	82	320	990				
5	490				21	63	250	760				
5	940	—	—	—	18	53	210	670	—	—	—	—
5	1,400				16	49	190	590				
6	500				33	130	400	1,000				
6	1,100	—	—	—	—	26	100	310	780	—	—	—
6	2,000					22	84	260	660			
6	2,900					20	77	240	600			
8	1,800	—	—	—	—		31	95	240	940	—	—
8	3,400						24	73	190	720		

(continued)

TABLE 916.1—continued
SIZE AND DEVELOPED LENGTH OF STACK VENTS AND VENT STACKS

DIAMETER OF SOIL OR WASTE STACK (inches)	TOTAL FIXTURE UNITS BEING VENTED (dfu)	MAXIMUM DEVELOPED LENGTH OF VENT (feet)[a] DIAMETER OF VENT (inches)										
		1¼	1½	2	2½	3	4	5	6	8	10	12
8	5,600						20	62	160	610		
8	7,600	—	—	—	—	—	18	56	140	560		—
10	4,000							31	78	310	960	
10	7,200							24	60	240	740	
10	11,000	—	—	—	—	—		20	51	200	630	—
10	15,000							18	46	180	570	
12	7,300								31	120	380	940
12	13,000	—	—	—	—	—	—		24	94	300	720
12	20,000								20	79	250	610
12	26,000								18	72	230	500
15	15,000	—	—	—	—	—	—	—		40	130	310
15	25,000									31	96	240
15	38,000									26	81	200
15	50,000	—	—	—	—	—	—	—	—	24	74	180

For SI: 1 inch = 25.4 mm, 1 foot = 304.8 mm.

a. The developed length shall be measured from the vent connection to the open air.

916.2 Vents other than stack vents or vent stacks. The diameter of individual vents, branch vents, circuit vents and relief vents shall be at least one-half the required diameter of the drain served. The required size of the drain shall be determined in accordance with Table 710.1(2). Vent pipes shall not be less than 1¼ inches (32 mm) in diameter. Vents exceeding 40 feet (12 192 mm) in developed length shall be increased by one nominal pipe size for the entire developed length of the vent pipe. Relief vents for soil and waste stacks in buildings having more than 10 branch intervals shall be sized in accordance with Section 914.2.

916.3 Developed length. The developed length of individual, branch, circuit and relief vents shall be measured from the farthest point of vent connection to the drainage system to the point of connection to the vent stack, stack vent or termination outside of the building.

916.4 Multiple branch vents. Where multiple branch vents are connected to a common branch vent, the common branch vent shall be sized in accordance with this section based on the size of the common horizontal drainage branch that is or would be required to serve the total drainage fixture unit (dfu) load being vented.

916.4.1 Branch vents exceeding 40 feet in developed length. Branch vents exceeding 40 feet (12 192 mm) in developed length shall be increased by one nominal size for the entire developed length of the vent pipe.

916.5 Sump vents. Sump vent sizes shall be determined in accordance with Sections 916.5.1 and 916.5.2.

916.5.1 Sewage pumps and sewage ejectors other than pneumatic. Drainage piping below sewer level shall be vented in a similar manner to that of a gravity system. Building sump vent sizes for sumps with sewage pumps or sewage ejectors, other than pneumatic, shall be determined in accordance with Table 916.5.1.

916.5.2 Pneumatic sewage ejectors. The air pressure relief pipe from a pneumatic sewage ejector shall be connected to an independent vent stack terminating as required for vent extensions through the roof. The relief pipe shall be sized to relieve air pressure inside the ejector to atmospheric pressure, but shall not be less than 1¼ inches (32 mm) in size.

SECTION 917
AIR ADMITTANCE VALVES

917.1 General. Vent systems utilizing air admittance valves shall comply with this section. Individual- and branch-type air admittance valves shall conform to ASSE 1051.

917.2 Installation. The valves shall be installed in accordance with the requirements of this section and the manufacturer's installation instructions. Air admittance valves shall be installed after the DWV testing required by Section 312.2 or 312.3 has been performed.

917.3 Where permitted. Individual, branch and circuit vents shall be permitted to terminate with a connection to an air admittance valve. The air admittance valve shall only vent fixtures that are on the same floor level and connect to a horizontal branch drain. The horizontal branch drain shall conform to Section 917.3.1 or Section 917.3.2.

TABLE 916.5.1
SIZE AND LENGTH OF SUMP VENTS

DISCHARGE CAPACITY OF PUMP (gpm)	MAXIMUM DEVELOPED LENGTH OF VENT (feet)[a]					
	Diameter of vent (inches)					
	$1^1/_4$	$1^1/_2$	2	$2^1/_2$	3	4
10	No limit[b]	No limit	No limit	No limit	No limit	No limit
20	270	No limit	No limit	No limit	No limit	No limit
40	72	160	No limit	No limit	No limit	No limit
60	31	75	270	No limit	No limit	No limit
80	16	41	150	380	No limit	No limit
100	10[c]	25	97	250	No limit	No limit
150	Not permitted	10[c]	44	110	370	No limit
200	Not permitted	Not permitted	20	60	210	No limit
250	Not permitted	Not permitted	10	36	132	No limit
300	Not permitted	Not permitted	10[c]	22	88	380
400	Not permitted	Not permitted	Not permitted	10[c]	44	210
500	Not permitted	Not permitted	Not permitted	Not permitted	24	130

For SI: 1 inch = 25.4 mm, 1 foot = 304.8 mm, 1 gallon per minute = 3.785 L/m.

a. Developed length plus an appropriate allowance for entrance losses and friction due to fittings, changes in direction and diameter. Suggested allowances shall be obtained from NSB Monograph 31 or other approved sources. An allowance of 50 percent of the developed length shall be assumed if a more precise value is not available.

b. Actual values greater than 500 feet.

c. Less than 10 feet.

917.3.1 Location of branch. The horizontal branch drain shall connect to the drainage stack or building drain a maximum of four branch intervals from the top of the stack.

917.3.2 Relief vent. The horizontal branch shall be provided with a relief vent that shall connect to a vent stack, or stack vent, or extend outdoors to the open air. The relief vent shall connect to the horizontal branch drain between the stack or building drain and the most downstream fixture drain connected to the horizontal branch drain. The relief vent shall be sized in accordance with Section 916.2 and installed in accordance with Section 905. The relief vent shall be permitted to serve as the vent for other fixtures.

917.4 Location. The air admittance valve shall be located a minimum of 4 inches (102 mm) above the horizontal branch drain or fixture drain being vented. The air admittance valve shall be located within the maximum developed length permitted for the vent. The air admittance valve shall be installed a minimum of 6 inches (152 mm) above insulation materials.

917.5 Access and ventilation. Access shall be provided to all air admittance valves. The valve shall be located within a ventilated space that allows air to enter the valve.

917.6 Size. The air admittance valve shall be rated in accordance with the standard for the size of the vent to which the valve is connected.

917.7 Vent required. Within each plumbing system, a minimum of one stack vent or vent stack shall extend outdoors to the open air.

917.8 Prohibited installations. Air admittance valves shall not be installed in nonneutralized special waste systems as described in Chapter 8. Valves shall not be located in spaces utilized as supply or return air plenums.

SECTION 918
ENGINEERED VENT SYSTEMS

918.1 General. Engineered vent systems shall comply with this section and the design, submittal, approval, inspection and testing requirements of Section 105.4.

918.2 Individual branch fixture and individual fixture header vents. The maximum developed length of individual fixture vents to vent branches and vent headers shall be determined in accordance with Table 918.2 for the minimum pipe diameters at the indicated vent airflow rates.

The individual vent airflow rate shall be determined in accordance with the following:

$$Q_{h,b} = N_{n,b} Q_v \qquad \text{(Equation 9-1)}$$

For SI: $Q_{h,b} = N_{n,b} Q_v (0.4719 \text{ L/s})$
where:

$N_{n,b}$ = Number of fixtures per header (or vent branch), total number of fixtures connected to vent stack.

$Q_{h,b}$ = Vent branch or vent header airflow rate (cfm).

Q_v = Total vent stack airflow rate (cfm).

$Q_v \text{ (gpm)} = 27.8 \, r_s^{\,2/3} (1 - r_s) D^{8/3}$

$Q_v \text{ (cfm)} = 0.134 \, Q_v \text{ (gpm)}$

where:

D = Drainage stack diameter (inches).

Q_w = Design discharge load (gpm).

r_s = Waste water flow area to total area.

$$= \frac{Q_w}{27.8 \, D^{8/3}}$$

Individual vent airflow rates are obtained by equally distributing $Q_{h,b}$ into one-half the total number of fixtures on the branch or header for more than two fixtures; for an odd number of total fixtures, decrease by one; for one fixture, apply the full value of $Q_{h,b}$.

Individual vent developed length shall be increased by 20 percent of the distance from the vent stack to the fixture vent connection on the vent branch or header.

SECTION 919
COMPUTERIZED VENT DESIGN

919.1 Design of vent system. The sizing, design and layout of the vent system shall be permitted to be determined by approved computer program design methods.

919.2 System capacity. The vent system shall be based on the air capacity requirements of the drainage system under a peak load condition.

TABLE 918.2
MINIMUM DIAMETER AND MAXIMUM LENGTH OF INDIVIDUAL BRANCH FIXTURE VENTS AND INDIVIDUAL
FIXTURE HEADER VENTS FOR SMOOTH PIPES

DIAMETER OF VENT PIPE (inches)	INDIVIDUAL VENT AIRFLOW RATE (cubic feet per minute)																			
	Maximum developed length of vent (feet)																			
	1	2	3	4	5	6	7	8	9	10	11	12	13	14	15	16	17	18	19	20
$^1/_2$	95	25	13	8	5	4	3	2	1	1	1	1	1	1	1	1	1	1	1	1
$^3/_4$	100	88	47	30	20	15	10	9	7	6	5	4	3	3	3	2	2	2	2	1
1	—	—	100	94	65	48	37	29	24	20	17	14	12	11	9	8	7	7	6	6
$1^1/_4$	—	—	—	—	—	—	—	100	87	73	62	53	46	40	36	32	29	26	23	21
$1^1/_2$	—	—	—	—	—	—	—	—	—	—	—	100	96	84	75	65	60	54	49	45
2	—	—	—	—	—	—	—	—	—	—	—	—	—	—	—	—	—	—	—	100

For SI: 1 inch = 25.4 mm, 1 cubic foot per minute = 0.4719 L/s, 1 foot = 304.8 mm.

CHAPTER 10

TRAPS, INTERCEPTORS AND SEPARATORS

SECTION 1001
GENERAL

1001.1 Scope. This chapter shall govern the material and installation of traps, interceptors and separators.

SECTION 1002
TRAP REQUIREMENTS

1002.1 Fixture traps. Each plumbing fixture shall be separately trapped by a water-seal trap, except as otherwise permitted by this code. The trap shall be placed as close as possible to the fixture outlet. The vertical distance from the fixture outlet to the trap weir shall not exceed 24 inches (610 mm). The distance of a clothes washer standpipe above a trap shall conform to Section 802.4. A fixture shall not be double trapped.

Exceptions:

1. This section shall not apply to fixtures with integral traps.

2. A combination plumbing fixture is permitted to be installed on one trap provided that one compartment is not more than 6 inches (152 mm) deeper than the other compartment and the waste outlets are not more than 30 inches (762 mm) apart.

3. A grease trap intended to serve as a fixture trap in accordance with the manufacturer's installation instructions shall be permitted to serve as the trap for a single fixture or a combination sink of not more than three compartments where the vertical distance from the fixture outlet to the inlet of the interceptor does not exceed 30 inches (762 mm), and the developed length of the waste pipe from the most upstream fixture outlet to the inlet of the interceptor does not exceed 60 inches (1524 mm).

1002.2 Design of traps. Fixture traps shall be self-scouring. Fixture traps shall not have interior partitions, except where such traps are integral with the fixture or where such traps are constructed of an approved material that is resistant to corrosion and degradation. Slip joints shall be made with an approved elastomeric gasket and shall be installed only on the trap inlet, trap outlet and within the trap seal.

1002.3 Prohibited traps. The following types of traps are prohibited:

1. Traps that depend on moving parts to maintain the seal.

2. Bell traps.

3. Crown-vented traps.

4. Traps not integral with a fixture and that depend on interior partitions for the seal, except those traps constructed of an approved material that is resistant to corrosion and degradation.

5. "S" traps.

6. Drum traps.

   Exception: Drum traps used as solids interceptors and drum traps serving chemical waste systems shall not be prohibited.

1002.4 Trap seals. Each fixture trap shall have a liquid seal of not less than 2 inches (51 mm) and not more than 4 inches (102 mm), or deeper for special designs relating to accessible fixtures. Where a trap seal is subject to loss by evaporation, a trap seal primer valve shall be installed. A trap seal primer valve shall conform to ASSE 1018 or ASSE 1044.

1002.5 Size of fixture traps. Fixture trap size shall be sufficient to drain the fixture rapidly and not less than the size indicated in Table 709.1. A trap shall not be larger than the drainage pipe into which the trap discharges.

1002.6 Building traps. Building (house) traps shall be prohibited, except where local conditions necessitate such traps. Building traps shall be provided with a cleanout and a relief vent or fresh air intake on the inlet side of the trap. The size of the relief vent or fresh air intake shall not be less than one-half the diameter of the drain to which the relief vent or air intake connects. Such relief vent or fresh air intake shall be carried above grade and shall be terminated in a screened outlet located outside the building.

1002.7 Trap setting and protection. Traps shall be set level with respect to the trap seal and, where necessary, shall be protected from freezing.

1002.8 Recess for trap connection. A recess provided for connection of the underground trap, such as one serving a bathtub in slab-type construction, shall have sides and a bottom of corrosion-resistant, insect- and verminproof construction.

1002.9 Acid-resisting traps. Where a vitrified clay or other brittleware, acid-resisting trap is installed underground, such trap shall be embedded in concrete extending 6 inches (152 mm) beyond the bottom and sides of the trap.

1002.10 Plumbing in mental health centers. In mental health centers, pipes and traps shall not be exposed.

SECTION 1003
INTERCEPTORS AND SEPARATORS

1003.1 Where required. Interceptors and separators shall be provided to prevent the discharge of oil, grease, sand and other substances harmful or hazardous to the building drainage system, the public sewer, or sewage treatment plant or processes.

1003.2 Approval. The size, type and location of each interceptor and of each separator shall be designed and installed in accordance with the manufacturer's instructions and the requirements of this section based on the anticipated conditions of use. Wastes that do not require treatment or separation shall not be discharged into any interceptor or separator.

1003.3 Grease traps and grease interceptors. Grease traps and grease interceptors shall comply with the requirements of Sections 1003.3.1 through 1003.3.4.2.

1003.3.1 Grease traps and grease interceptors required. A grease trap or grease interceptor shall be required to receive the drainage from fixtures and equipment with grease-laden waste located in food preparation areas, such as in restaurants, hotel kitchens, hospitals, school kitchens, bars, factory cafeterias, or restaurants and clubs.

1003.3.2 Food waste grinders. Where food waste grinders connect to grease traps, a solids interceptor shall separate the discharge before connecting to the grease trap. Solids interceptors and grease interceptors shall be sized and rated for the discharge of the food waste grinder.

1003.3.3 Grease trap and grease interceptor not required. A grease trap or a grease interceptor shall not be required for individual dwelling units or any private living quarters.

1003.3.4 Grease traps and grease interceptors. Grease traps and grease interceptors shall conform to PDI G101, ASME A112.14.3 or ASME A112.14.4 and shall be installed in accordance with the manufacturer's instructions.

1003.3.4.1 Grease trap capacity. Grease traps shall have the grease retention capacity indicated in Table 1003.3.4.1 for the flow-through rates indicated.

TABLE 1003.3.4.1
CAPACITY OF GREASE TRAPS

TOTAL FLOW-THROUGH RATING (gpm)	GREASE RETENTION CAPACITY (pounds)
4	8
6	12
7	14
9	18
10	20
12	24
14	28
15	30
18	36
20	40
25	50
35	70
50	100

For SI: 1 gallon per minute = 3.785 L/m, 1 pound = 0.454 kg.

1003.3.4.2 Rate of flow controls. Grease traps shall be equipped with devices to control the rate of water flow so that the water flow does not exceed the rated flow. The flow-control device shall be vented and terminate not less than 6 inches (152 mm) above the flood rim level or be installed in accordance with the manufacturer's instructions.

1003.4 Oil separators required. At repair garages, carwashing facilities with engine or undercarriage cleaning capability and at factories where oily and flammable liquid wastes are produced, separators shall be installed into which all oil-bearing, grease-bearing or flammable wastes shall be discharged before emptying in the building drainage system or other point of disposal.

1003.4.1 Separation of liquids. A mixture of treated or untreated light and heavy liquids with various specific gravities shall be separated in an approved receptacle.

1003.4.2 Oil separator design. Oil separators shall be designed in accordance with Sections 1003.4.2.1 and 1003.4.2.2.

1003.4.2.1 General design requirements. Oil separators shall have a depth of not less than 2 feet (610 mm) below the invert of the discharge drain. The outlet opening of the separator shall have not less than an 18-inch (457 mm) water seal.

1003.4.2.2 Garages and service stations. Where automobiles are serviced, greased, repaired or washed or where gasoline is dispensed, oil separators shall have a minimum capacity of 6 cubic feet (0.168 m3) for the first 100 square feet (9.3 m2) of area to be drained, plus 1 cubic foot(0.28 m3) for each additional 100 square feet (9.3 m2) of area to be drained into the separator. Parking garages in which servicing, repairing or washing is not conducted, and in which gasoline is not dispensed, shall not require a separator. Areas of commercial garages utilized only for storage of automobiles are not required to be drained through a separator.

1003.5 Sand interceptors in commercial establishments. Sand and similar interceptors for heavy solids shall be designed and located so as to be provided with ready access for cleaning, and shall have a water seal of not less than 6 inches (152 mm).

1003.6 Laundries. Commercial laundries shall be equipped with an interceptor with a wire basket or similar device, removable for cleaning, that prevents passage into the drainage system of solids 0.5 inch (12.7 mm) or larger in size, string, rags, buttons or other materials detrimental to the public sewage system.

1003.7 Bottling establishments. Bottling plants shall discharge process wastes into an interceptor that will provide for the separation of broken glass or other solids before discharging waste into the drainage system.

1003.8 Slaughterhouses. Slaughtering room and dressing room drains shall be equipped with approved separators. The separator shall prevent the discharge into the drainage system of feathers, entrails and other materials that cause clogging.

1003.9 Venting of interceptors and separators. Interceptors and separators shall be designed so as not to become air bound where tight covers are utilized. Each interceptor or separator shall be vented where subject to a loss of trap seal.

1003.10 Access and maintenance of interceptors and separators. Access shall be provided to each interceptor and separator for service and maintenance. Interceptors and separators shall be maintained by periodic removal of accumulated grease, scum, oil, or other floating substances and solids deposited in the interceptor or separator.

SECTION 1004
MATERIALS, JOINTS AND CONNECTIONS

1004.1 General. The materials and methods utilized for the construction and installation of traps, interceptors and separators shall comply with this chapter and the applicable provisions of Chapters 4 and 7. The fittings shall not have ledges, shoulders or reductions capable of retarding or obstructing flow of the piping.

CHAPTER 11

STORM DRAINAGE

SECTION 1101
GENERAL

1101.1 Scope. The provisions of this chapter shall govern the materials, design, construction and installation of storm drainage.

1101.2 Where required. All roofs, paved areas, yards, courts and courtyards shall drain into a separate storm sewer system, or a combined sewer system, or to an approved place of disposal. For one- and two-family dwellings, and where approved, storm water is permitted to discharge onto flat areas, such as streets or lawns, provided that the storm water flows away from the building.

1101.3 Prohibited drainage. Storm water shall not be drained into sewers intended for sewage only.

1101.4 Tests. The conductors and the building storm drain shall be tested in accordance with Section 312.

1101.5 Continuous flow. The size of a drainage pipe shall not be reduced in the direction of flow.

1101.6 Fittings and connections. All connections and changes in direction of the storm drainage system shall be made with approved drainage-type fittings in accordance with Table 706.3. The fittings shall not obstruct or retard flow in the system.

1101.7 Roof design. Roofs shall be designed for the maximum possible depth of water that will pond thereon as determined by the relative levels of roof deck and overflow weirs, scuppers, edges or serviceable drains in combination with the deflected structural elements. In determining the maximum possible depth of water, all primary roof drainage means shall be assumed to be blocked.

1101.8 Cleanouts required. Cleanouts shall be installed in the storm drainage system and shall comply with the provisions of this code for sanitary drainage pipe cleanouts.

Exception: Subsurface drainage system.

1101.9 Backwater valves. Backwater valves installed in a storm drainage system shall conform to Section 715.

SECTION 1102
MATERIALS

1102.1 General. The materials and methods utilized for the construction and installation of storm drainage systems shall comply with this section and the applicable provisions of Chapter 7.

1102.2 Inside storm drainage conductors. Inside storm drainage conductors installed above ground shall conform to one of the standards listed in Table 702.1.

1102.3 Underground building storm drain pipe. Underground building storm drain pipe shall conform to one of the standards listed in Table 702.2.

1102.4 Building storm sewer pipe. Building storm sewer pipe shall conform to one of the standards listed in Table 1102.4.

TABLE 1102.4
BUILDING STORM SEWER PIPE

MATERIAL	STANDARD
Acrylonitrile butadiene styrene (ABS) plastic pipe	ASTM D 2661; ASTM D 2751; ASTM F 628
Asbestos-cement pipe	ASTM C 428
Cast-iron pipe	ASTM A 74; ASTM A 888; CISPI 301
Concrete pipe	ASTM C 14; ASTM C 76; CAN/CSA A257.1M; CAN/CSA A257.2M
Copper or copper-alloy tubing (Type K, L, M or DWV)	ASTM B 75; ASTM B 88; ASTM B 251; ASTM B 306
Polyvinyl chloride (PVC) plastic pipe (Type DWV, SDR26, SDR35, SDR41, PS50 or PS100)	ASTM D 2665; ASTM D 3034; ASTM F 891; CSA-B182.2; CAN/CSA-B182.4
Vitrified clay pipe	ASTM C 4; ASTM C 700
Stainless steel drainage Systems, Type 316L	ASME A112.3.1

1102.5 Subsoil drain pipe. Subsoil drains shall be open-jointed, horizontally split or perforated pipe conforming to one of the standards listed in Table 1102.5.

TABLE 1102.5
SUBSOIL DRAIN PIPE

MATERIAL	STANDARD
Asbestos-cement pipe	ASTM C 508
Cast-iron pipe	ASTM A 74; ASTM A 888; CISPI 301
Polyethylene (PE) plastic pipe	ASTM F 405
Polyvinyl chloride (PVC) Plastic pipe (type sewer pipe, PS25, PS50 or PS100)	ASTM D 2729; ASTM F 891; CSA-B 182.2; CAN/CSA-B182.4
Vitrified clay pipe	ASTM C 4; ASTM C 700
Stainless steel drainage Systems, Type 316L	ASME A112.3.1

1102.6 Roof drains. Roof drains shall conform to ASME A112.21.2M or ASME A112.3.1.

1102.7 Fittings. Pipe fittings shall be approved for installation with the piping material installed, and shall conform to the respective pipe standards or one of the standards listed in Table 1102.7. The fittings shall not have ledges, shoulders or reductions capable of retarding or obstructing flow in the piping. Threaded drainage pipe fittings shall be of the recessed drainage type.

TABLE 1102.7
PIPE FITTINGS

MATERIAL	STANDARD
Acrylonitrile butadiene styrene (ABS) plastic	ASTM D 2468; ASTM D 2661
Cast-iron	ASME B16.4; ASME B16.12; ASTM A 888; CISPI 301; ASTM A 74
Chlorinated polyvinyl chloride (CPVC) plastic	ASTM F 437; ASTM F 438; ASTM F 439
Copper or copper alloy	ASME B16.15; ASME B16.18; ASME B16.22; ASME B16.23; ASME B16.26; ASME B16.29
Gray iron and ductile iron	AWWA C110
Malleable iron	ASME B16.3
Plastic, general	ASTM F 409
Polyethylene (PE) plastic	ASTM D 2609
Polyvinyl chloride (PVC) plastic	ASTM D 2464; ASTM D 2466; ASTM D 2467; CSA-B137.2; ASTM D 2665; ASTM F 1866
Steel	ASME B16.9; ASME B16.11; ASME B16.28
Stainless steel drainage Systems, Type 316L	ASME A112.3.1

SECTION 1103
TRAPS

1103.1 Main trap. Leaders and storm drains connected to a combined sewer shall be trapped. Individual storm water traps shall be installed on the storm water drain branch serving each conductor, or a single trap shall be installed in the main storm drain just before its connection with the combined building sewer or the public sewer.

1103.2 Material. Storm water traps shall be of the same material as the piping system to which they are attached.

1103.3 Size. Traps for individual conductors shall be the same size as the horizontal drain to which they are connected.

1103.4 Cleanout. An accessible cleanout shall be installed on the building side of the trap.

SECTION 1104
CONDUCTORS AND CONNECTIONS

1104.1 Prohibited use. Conductor pipes shall not be used as soil, waste or vent pipes, and soil, waste or vent pipes shall not be used as conductors.

1104.2 Combining storm with sanitary drainage. The sanitary and storm drainage systems of a structure shall be entirely separate except where combined sewer systems are utilized. Where a combined sewer is utilized, the building storm drain shall be connected in the same horizontal plane through a single-wye fitting to the combined sewer at least 10 feet(3048 mm) downstream from any soil stack.

1104.3 Floor drains. Floor drains shall not be connected to a storm drain.

SECTION 1105
ROOF DRAINS

1105.1 Strainers. Roof drains shall have strainers extending not less than 4 inches (102 mm) above the surface of the roof immediately adjacent to the roof drain. Strainers shall have an available inlet area, above roof level, of not less than one and one-half times the area of the conductor or leader to which the drain is connected.

1105.2 Flat decks. Roof drain strainers for use on sun decks, parking decks and similar areas that are normally serviced and maintained shall comply with Section 1105.1 or shall be of the flat-surface type, installed level with the deck, with an available inlet area not less than two times the area of the conductor or leader to which the drain is connected.

1105.3 Roof drain flashings. The connection between roofs and roof drains which pass through the roof and into the interior of the building shall be made water tight by the use of approved flashing material.

SECTION 1106
SIZE OF CONDUCTORS, LEADERS
AND STORM DRAINS

1106.1 General. The size of the vertical conductors and leaders, building storm drains, building storm sewers, and any horizontal branches of such drains or sewers shall be based on the 100-year hourly rainfall rate indicated in Figure 1106.1 or on other rainfall rates determined from approved local weather data.

1106.2 Vertical conductors and leaders. Vertical conductors and leaders shall be sized for the maximum projected roof area, in accordance with Table 1106.2.

1106.3 Building storm drains and sewers. The size of the building storm drain, building storm sewer and their horizontal branches having a slope of one-half unit or less vertical in 12 units horizontal (4-percent slope) shall be based on the maximum projected roof area in accordance with Table 1106.3. The minimum slope of horizontal branches shall be one-eighth unit vertical in 12 units horizontal (1-percent slope) unless otherwise approved.

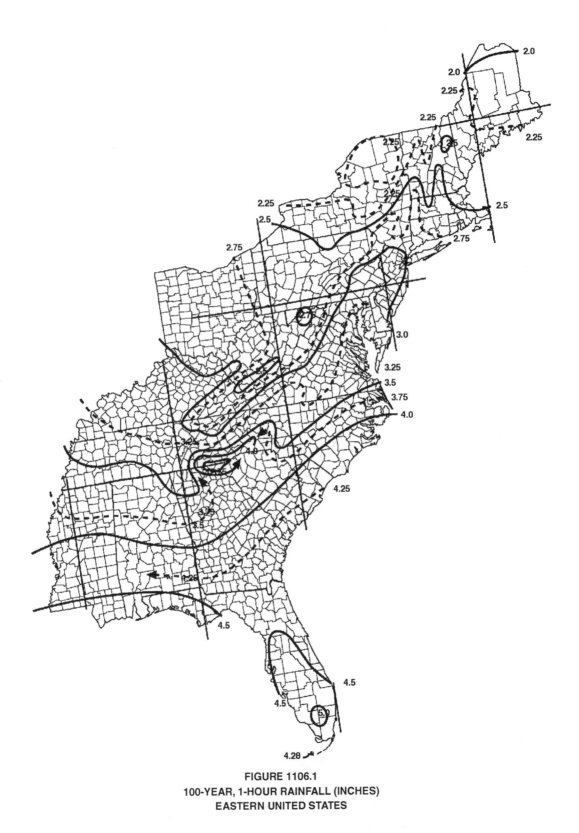

FIGURE 1106.1
100-YEAR, 1-HOUR RAINFALL (INCHES)
EASTERN UNITED STATES

For SI: 1 inch = 25.4 mm.

Source: National Weather Service, National Oceanic and Atmospheric Administration, Washington D.C.

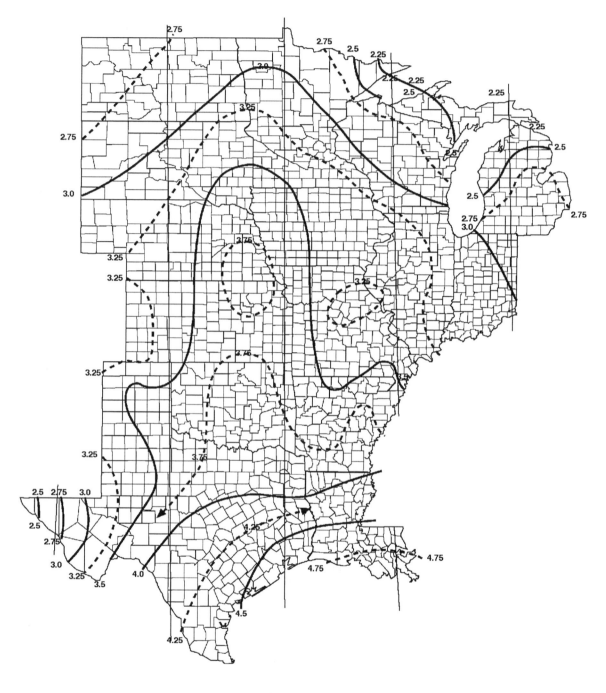

FIGURE 1106.1—continued
100-YEAR, 1-HOUR RAINFALL (INCHES)
CENTRAL UNITED STATES

For SI: 1 inch = 25.4 mm.

Source: National Weather Service, National Oceanic and Atmospheric Administration, Washington D.C.

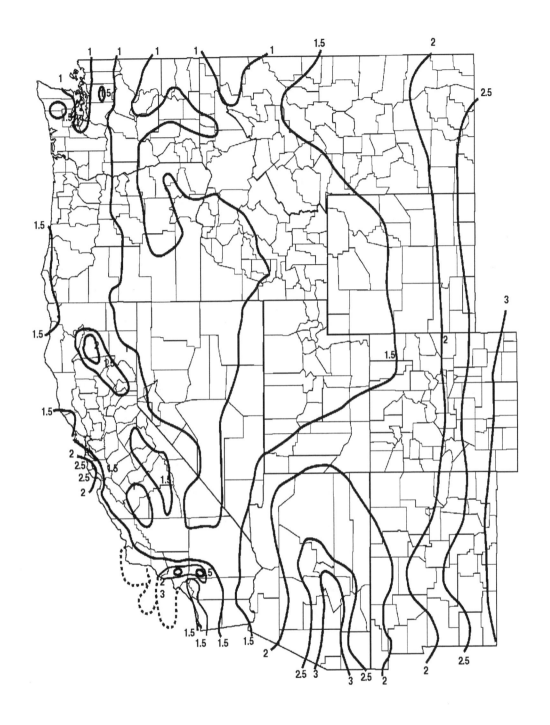

FIGURE 1106.1—continued
100-YEAR, 1-HOUR RAINFALL (INCHES)
WESTERN UNITED STATES

For SI: 1 inch = 25.4 mm.

Source: National Weather Service, National Oceanic and Atmospheric Administration, Washington D.C.

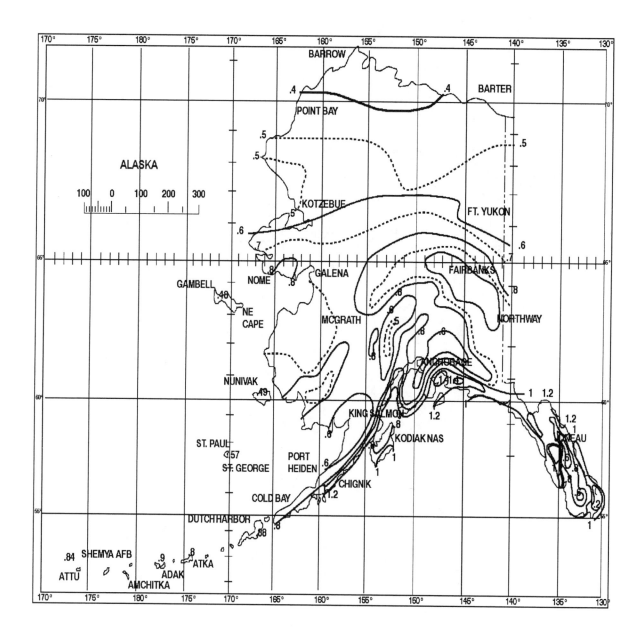

FIGURE 1106.1—continued
100-YEAR, 1-HOUR RAINFALL (INCHES)
ALASKA

For SI: 1 inch = 25.4 mm.

Source: National Weather Service, National Oceanic and Atmospheric Administration, Washington D.C.

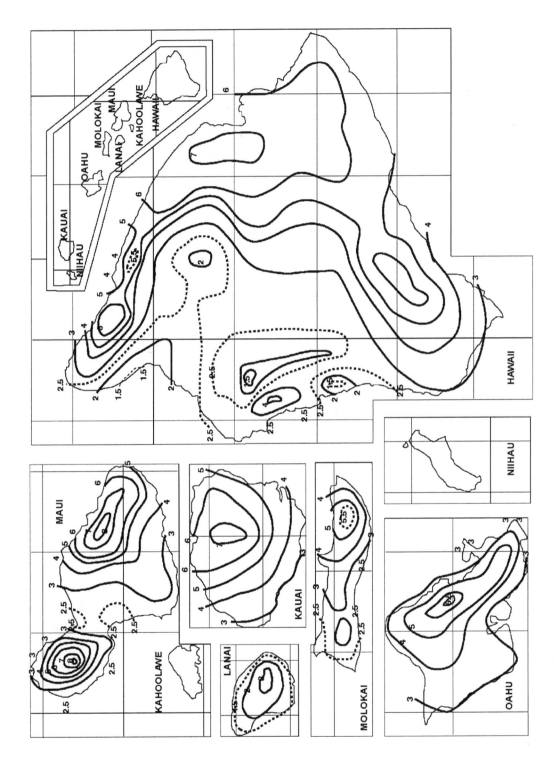

FIGURE 1106.1—continued
100-YEAR, 1-HOUR RAINFALL (INCHES)
HAWAII

For SI: 1 inch = 25.4 mm.

Source: National Weather Service, National Oceanic and Atmospheric Administration, Washington D.C.

TABLE 1106.2
SIZE OF VERTICAL CONDUCTORS AND LEADERS

DIAMETER OF OF LEADER (inches)[a]	HORIZONTALLY PROJECTED ROOF AREA (square feet)											
	Rainfall rate (inches per hour)											
	1	2	3	4	5	6	7	8	9	10	11	12
2	2,880	1,440	960	720	575	480	410	360	320	290	260	240
3	8,800	4,400	2,930	2,200	1,760	1,470	1,260	1,100	980	880	800	730
4	18,400	9,200	6,130	4,600	3,680	3,070	2,630	2,300	2,045	1,840	1,675	1,530
5	34,600	17,300	11,530	8,650	6,920	5,765	4,945	4,325	3,845	3,460	3,145	2,880
6	54,000	27,000	17,995	13,500	10,800	9,000	7,715	6,750	6,000	5,400	4,910	4,500
8	116,000	58,000	38,660	29,000	23,200	19,315	16,570	14,500	12,890	11,600	10,545	9,600

For SI: 1 inch = 25.4 mm, 1 square foot = 0.0929 m^2.

a. Sizes indicated are the diameter of circular piping. This table is applicable to piping of other shapes provided the cross-sectional shape fully encloses a circle of the diameter indicated in this table.

TABLE 1106.3
SIZE OF HORIZONTAL STORM DRAINGE PIPING

SIZE OF HORIZONTAL PIPING (inches)	HORIZONTALLY PROJECTED ROOF AREA (square feet)					
	Rainfall rate (inches per hour)					
	1	2	3	4	5	6
$^1/_8$ unit vertical in 12 units horizontal (1-percent slope)						
3	3,288	1,644	1,096	822	657	548
4	7,520	3,760	2,506	1,800	1,504	1,253
5	13,360	6,680	4,453	3,340	2,672	2,227
6	21,400	10,700	7,133	5,350	4,280	3,566
8	46,000	23,000	15,330	11,500	9,200	7,600
10	82,800	41,400	27,600	20,700	16,580	13,800
12	133,200	66,600	44,400	33,300	26,650	22,200
15	218,000	109,000	72,800	59,500	47,600	39,650
$^1/_4$ unit vertical in 12 units horizontal (2-percent slope)						
3	4,640	2,320	1,546	1,160	928	773
4	10,600	5,300	3,533	2,650	2,120	1,766
5	18,880	9,440	6,293	4,720	3,776	3,146
6	30,200	15,100	10,066	7,550	6,040	5,033
8	65,200	32,600	21,733	16,300	13,040	10,866
10	116,800	58,400	38,950	29,200	23,350	19,450
12	188,000	94,000	62,600	47,000	37,600	31,350
15	336,000	168,000	112,000	84,000	67,250	56,000
$^1/_2$ unit vertical in 12 units horizontal (4-percent slope)						
3	6,576	3,288	2,295	1,644	1,310	1,096
4	15,040	7,520	5,010	3,760	3,010	2,500
5	26,720	13,360	8,900	6,680	5,320	4,450
6	42,800	21,400	13,700	10,700	8,580	7,140
8	92,000	46,000	30,650	23,000	18,400	15,320
10	171,600	85,800	55,200	41,400	33,150	27,600
12	266,400	133,200	88,800	66,600	53,200	44,400
15	476,000	238,000	158,800	119,000	95,300	79,250

For SI: 1 inch = 25.4 mm, 1 square foot = 0.0929 m^2.

1106.4 Vertical walls. In sizing roof drains and storm drainage piping, one-half of the area of any vertical wall that diverts rainwater to the roof shall be added to the projected roof area for inclusion in calculating the required size of vertical conductors, leaders and horizontal storm drainage piping.

1106.5 Parapet wall scupper location. Parapet wall roof drainage scupper and overflow scupper location shall comply with the requirements of the *International Building Code*.

1106.6 Size of roof gutters. The size of semicircular gutters shall be based on the maximum projected roof area in accordance with Table 1106.6.

SECTION 1107
SECONDARY (EMERGENCY) ROOF DRAINS

1107.1 Secondary drainage required. Secondary (emergency) roof drains or scuppers shall be provided where the roof perimeter construction extends above the roof in such a manner that water will be entrapped if the primary drains allow buildup for any reason.

TABLE 1106.6
SIZE OF SEMICIRCULAR ROOF GUTTERS

DIAMETER OF GUTTERS (inches)	HORIZONTALLY PROJECTED ROOF AREA (square feet)					
	Rainfall rate (inches per hour)					
	1	2	3	4	5	6
$^1/_{16}$ unit vertical in 12 units horizontal (0.5-percent slope)						
3	680	340	226	170	136	113
4	1,440	720	480	360	288	240
5	2,500	1,250	834	625	500	416
6	3,840	1,920	1,280	960	768	640
7	5,520	2,760	1,840	1,380	1,100	918
8	7,960	3,980	2,655	1,990	1,590	1,325
10	14,400	7,200	4,800	3,600	2,880	2,400
$^1/_8$ unit vertical 12 units horizontal (1-percent slope)						
3	960	480	320	240	192	160
4	2,040	1,020	681	510	408	340
5	3,520	1,760	1,172	880	704	587
6	5,440	2,720	1,815	1,360	1,085	905
7	7,800	3,900	2,600	1,950	1,560	1,300
8	11,200	5,600	3,740	2,800	2,240	1,870
10	20,400	10,200	6,800	5,100	4,080	3,400
¼ unit vertical in 12 units horizontal (2-percent slope)						
3	1,360	680	454	340	272	226
4	2,880	1,440	960	720	576	480
5	5,000	2,500	1,668	1,250	1,000	834
6	7,680	3,840	2,560	1,920	1,536	1,280
7	11,040	5,520	3,860	2,760	2,205	1,840
8	15,920	7,960	5,310	3,980	3,180	2,655
10	28,800	14,400	9,600	7,200	5,750	4,800
½ unit vertical in 12 units horizontal (4-percent)						
3	1,920	960	640	480	384	320
4	4,080	2,040	1,360	1,020	816	680
5	7,080	3,540	2,360	1,770	1,415	1,180
6	11,080	5,540	3,695	2,770	2,220	1,850
7	15,600	7,800	5,200	3,900	3,120	2,600
8	22,400	11,200	7,460	5,600	4,480	3,730
10	40,000	20,000	13,330	10,000	8,000	6,660

For SI: 1 inch = 25.4 mm, 1 square foot = 0.0929 m².

1107.2 Separate systems required. Secondary roof drain systems shall have the end point of discharge separate from the primary system. Discharge shall be above grade, in a location which would normally be observed by the building occupants or maintenance personnel.

1107.3 Sizing of secondary drains. Secondary (emergency) roof drain systems shall be sized in accordance with Section 1106 based on the rainfall rate for which the primary system is sized in Tables 1106.2, 1106.3 and 1106.6. Scuppers shall be sized to prevent the depth of ponding water from exceeding that for which the roof was designed as determined by Section 1101.7. Scuppers shall not have an opening dimension of less than 4 inches (102 mm). The flow through the primary system shall not be considered when sizing the secondary roof drain system.

SECTION 1108
COMBINED SANITARY AND STORM SYSTEM

1108.1 Size of combined drains and sewers. The size of a combination sanitary and storm drain or sewer shall be computed in accordance with the method in Section 1106.3. The fixture units shall be converted into an equivalent projected roof or paved area. Where the total fixture load on the combined drain is less than or equal to 256 fixture units, the equivalent drainage area in horizontal projection shall be taken as 4,000 square feet (372 m²). Where the total fixture load exceeds 256 fixture units, each additional fixture unit shall be considered the equivalent of 15.6 square feet (1.5 m²) of drainage area. These values are based on a rainfall rate of 1 inch (25 mm) per hour.

SECTION 1109
VALUES FOR CONTINUOUS FLOW

1109.1 Equivalent roof area. Where there is a continuous or semicontinuous discharge into the building storm drain or building storm sewer, such as from a pump, ejector, air conditioning plant or similar device, each gallon per minute (L/m) of such discharge shall be computed as being equivalent to 96 square feet (9 m²) of roof area, based on a rainfall rate of 1 inch (25.4 mm) per hour.

SECTION 1110
CONTROLLED FLOW ROOF DRAIN SYSTEMS

1110.1 General. The roof of a structure shall be designed for the storage of water where the storm drainage system is engineered for controlled flow. The controlled flow roof drain system shall be an engineered system in accordance with this section and the design, submittal, approval, inspection and testing requirements of Section 105.4. The controlled flow system shall be designed based on the required rainfall rate in accordance with Section 1106.1.

1110.2 Control devices. The control devices shall be installed so that the rate of discharge of water per minute shall not exceed the values for continuous flow as indicated in Section 1109.1.

1110.3 Installation. Runoff control shall be by control devices. Control devices shall be protected by strainers.

1110.4 Minimum number of roof drains. Not less than two roof drains shall be installed in roof areas 10,000 square feet (930 m²) or less and not less than four roof drains shall be installed in roofs over 10,000 square feet (930 m²) in area.

SECTION 1111
SUBSOIL DRAINS

1111.1 Subsoil drains. Subsoil drains shall be open-jointed, horizontally split or perforated pipe conforming to one of the standards listed in Table 1102.5. Such drains shall not be less than 4 inches (102 mm) in diameter. Where the building is subject to backwater, the subsoil drain shall be protected by an accessibly located backwater valve. Subsoil drains shall discharge to a trapped area drain, sump, dry well or approved location above ground. The subsoil sump shall not be required to have either a gas-tight cover or a vent. The sump and pumping system shall comply with Section 1113.1.

SECTION 1112
BUILDING SUBDRAINS

1112.1 Building subdrains. Building subdrains located below the public sewer level shall discharge into a sump or receiving tank, the contents of which shall be automatically lifted and discharged into the drainage system as required for building sumps. The sump and pumping equipment shall comply with Section 1113.1.

SECTION 1113
SUMPS AND PUMPING SYSTEMS

1113.1 Pumping system. The sump pump, pit and discharge piping shall conform to Sections 1113.1.1 through 1113.1.4.

1113.1.1 Pump capacity and head. The sump pump shall be of a capacity and head appropriate to anticipated use requirements.

1113.1.2 Sump pit. The sump pit shall not be less than 18 inches (457 mm) in diameter and 24 inches (610 mm) deep, unless otherwise approved. The pit shall be accessible and located such that all drainage flows into the pit by gravity. The sump pit shall be constructed of tile, steel, plastic, cast-iron, concrete or other approved material, with a removable cover adequate to support anticipated loads in the area of use. The pit floor shall be solid and provide permanent support for the pump.

1113.1.3 Electrical. Electrical service outlets, when required, shall meet the requirements of the ICC *Electrical Code*.

1113.1.4 Piping. Discharge piping shall meet the requirements of Section 1102.2, 1102.3 or 1102.4 and shall include a gate valve and a full flow check valve. Pipe and fittings shall be the same size as, or larger than, pump discharge tapping.

> **Exception:** In one- and two-family dwellings, only a check valve shall be required, located on the discharge piping from the pump or ejector.

CHAPTER 12

SPECIAL PIPING AND STORAGE SYSTEMS

SECTION 1201
GENERAL

1201.1 Scope. The provisions of this chapter shall govern the design and installation of piping and storage systems for non-flammable medical gas systems and nonmedical oxygen systems. All maintenance and operations of such systems shall be in accordance with the *International Fire Code.*

SECTION 1202
MEDICAL GASES

[F] 1202.1 Nonflammable medical gases. Nonflammable medical gas systems, inhalation anesthetic systems and vacuum piping systems shall be designed and installed in accordance with NFPA 99C.

Exceptions:

1. This section shall not apply to portable systems or cylinder storage.

2. Vacuum system exhaust shall comply with the *International Mechanical Code.*

SECTION 1203
OXYGEN SYSTEMS

[F] 1203.1 Design and installation. Nonmedical oxygen systems shall be designed and installed in accordance with NFPA 50 and NFPA 51.

CHAPTER 13

REFERENCED STANDARDS

This chapter lists the standards that are referenced in various sections of this document. The standards are listed herein by the promulgating agency of the standard, the standard identification, the effective date and title, and the section or sections of this document that reference the standard. The application of the referenced standards shall be as specified in Section 102.8.

ANSI

American National Standards Institute
25 West 43rd Street, Fourth Floor
New York, NY 10036

Standard Reference Number	Title	Referenced in code section number
Z4.3—95	Minimum Requirements for Nonsewered Waste-Disposal Systems	311.1
Z21.22—99	Relief Valves for Hot Water Supply Systems	504.2, 504.5
Z124.1—95	Plastic Bathtub Units	407.1
Z124.2—95	Plastic Shower Receptors and Shower Stalls	417.1
Z124.3—95	Plastic Lavatories	416.1, 416.2
Z124.4—96	Plastic Water Closet Bowls and Tanks	420.1
Z124.6—97	Plastic Sinks	415.1, 418.1

ARI

Air-Conditioning & Refrigeration Institute
4100 North Fairfax Drive, Suite 200
Arlington, VA 22203

Standard reference number	Title	Referenced in code section number
1010—94	Self-Contained, Mechanically-Refrigerated Drinking-Water Coolers	410.1

ASME

American Society of Mechanical Engineers
Three Park Avenue
New York, NY 10016-5990

Standard Reference Number	Title	Referenced in code section number
A112.1.2—1991(R1998)	Air Gaps in Plumbing Systems	Table 608.1
A112.1.3—2000	Air Gap Fittings for Use with Plumbing Fixtures, Appliances and Appurtenances	608.13.1, Table 608.1
A112.3.1—1993	Performance Standard and Installation Procedures for Stainless Steel Drainage Systems or Sanitary, Storm and Chemical Applications, Above and Below Ground	412.1, Table 702.1, Table 702.2, Table 702.3, Table 702.4, 708.2, Table 1102.4, Table 1102.5, 1102.6, Table 1102.7
A112.3.4—2000	Macerating Toilet Systems and Related Components	712.4.1
A112.4.1—1993(R1998)	Water Heater Relief Valve Drain Tubes	504.6.2
A112.4.3—1999	Plastic Fittings for Connecting Water Closets to the Sanitary Drainage System	405.4
A112.6.1M—1997	Floor-Affixed Supports for Off-the-Floor Plumbing Fixtures for Public Use	405.4.3
A112.6.2—2000	Framing-Affixed Supports for Off-the-Floor Water Closets with Concealed Tanks	405.4.3
A112.6.3—2001	Floor and Trench Drains	412.1
A112.6.7—2001	Enameled and Epoxy-coated Cast-iron and PVC plastic Sanitary Floor Sinks	427.1
A112.14.1—1975(R1998)	Backwater Valves	715.2
A112.14.3—2000	Grease Interceptors	1003.3.4
A112.14.4—2001	Grease Removal Devices	1003.3.4
A112.18.1—2000	Plumbing Fixture Fittings	424.1
A112.18.3M—1996	Performance Requirements for Backflow Protection Devices and Systems in Plumbing Fixture Fittings	424.5
A112.18.6—1999	Flexible Water Connectors	605.6
A112.18.7—1999	Deck mounted Bath/Shower Transfer Valves with Integral Backflow Protection	424.6
A112.19.1M—1994(R1999)	Enameled Cast Iron Plumbing Fixtures	407.1, 410.1, 415.1, 416.1, 418.1
A112.19.2M—1998	Vitreous China Plumbing Fixtures	401.2, 405.9, 408.1, 410.1, 416.1, 418.1, 419.1, 420.1

ASME—continued

A112.19.3M—1987(R1996)	Stainless Steel Plumbing Fixtures (Designed for Residential Use)	405.9, 415.1, 416.1, 418.1
A112.19.4M—1994(R1999)	Porcelain Enameled Formed Steel Plumbing Fixtures	407.1, 416.1, 418.1
A112.19.5—1999	Trim for Water-Closet Bowls, Tanks, and Urinals	425.4
A112.19.6—1995	Hydraulic Performance Requirements for Water Closets and Urinals	419.1, 420.1
A112.19.7M—1995	Whirlpool Bathtub Appliances	421.1
A112.19.8M—1987(R1996)	Suction Fittings for Use in Swimming Pools, Wading Pools, Spas, Hot Tubs, and Whirlpool Bathtub Appliances	421.4
A112.19.9M—1998	Non-Vitreous Ceramic Plumbing Fixtures	407.1, 408.1, 410.1, 415.1, 416.1, 417.1, 418.1, 420.1
A112.19.12—2000	Wall Mounted and Pedestal Mounted, Adjustable and Pivoting Lavatory and Sink Carrier Systems	416.4, 418.3
A112.19.13—2001	Electrohydraulic Water Closets	420.1
A112.19.15—2001	Bathtub/Whirlpool Bathtubs with Pressure Sealed Doors	407.4, 421.5
A112.21.2M—1983	Roof Drains	1102.6
A112.36.2M—1991(R1998)	Cleanouts	708.2
B1.20.1—1983(R1999)	Pipe Threads, General Purpose (inch)	605.10.3, 605.12.3, 605.14.4, 605.16.3, 605.18.1, 705, 705.2.3, 705.4.3
B16.3—1999	Malleable Iron Threaded Fittings Classes 150 and 300	Table 605.5, Table 702.4, Table 1102.7
B16.4—1998	Gray Iron Threaded Fittings Classes 125 and 250	Table 605.5, Table 702.4, Table 1102.7
B16.9—1993	Factory-Made Wrought Steel Buttwelding Fittings	Table 605.5, Table 702.4, Table 1102.7
B16.11—1996	Forged Fittings, Socket-Welding and Threaded	Table 605.5, Table 702.4, Table 1102.7
B16.12—1998	Cast-Iron Threaded Drainage Fittings	Table 605.5, Table 702.4, Table 1102.7
B16.15—1985(R1994)	Cast Bronze Threaded Fittings	Table 605.5, Table 702.4, Table 1102.7
B16.18—1984(R1994)	Cast Copper Alloy Solder Joint Pressure Fittings	Table 605.5, Table 702.4, Table 1102.7
B16.22—1995	Wrought Copper and Copper Alloy Solder Joint Pressure Fittings	Table 605.5, Table 702.4, Table 1102.7
B16.23—1992	Cast Copper Alloy Solder Joint Drainage Fittings DWV	Table 605.5, Table 702.4, Table 1102.7
B16.26—1988	Cast Copper Alloy Fittings for Flared Copper Tubes	Table 605.5, Table 702.4, Table 1102.7
B16.28—1994	Wrought Steel Buttwelding Short Radius Elbows and Returns	Table 605.5, Table 702.4, Table 1102.7
B16.29—1994	Wrought Copper and Wrought Copper Alloy Solder Joint Drainage Fittings—DWV	Table 605.5, Table 702.4, Table 1102.7

ASSE

American Society of Sanitary Engineering
901 Canterbury Road, Suite A
Westlake, OH 44145

Standard Reference Number	Title	Referenced in code section number
1001—90	Performance Requirements for Pipe Applied Atmospheric Type Vacuum Breakers	425.2, Table 608.1, 608.13.6
1002—99	Performance Requirements for Water Closet Flush Tank Ball Cocks	425.3.1, Table 608.1
1003—95	Performance Requirements for Water Pressure Reducing Valves	604.8
1004—90	Performance Requirements for Commercial Dishwashing Machines	409.1
1005—99	Performance Requirements for Water Heater Drain Valves	501.3
1006—89	Performance Requirements for Residential Use (Household) Dishwashers	409.1
1007—92	Performance Requirements for Home Laundry Equipment	406.1, 406.2
1008—89	Performance Requirements for Household Food Waste Disposer Units	413.1
1009—90	Performance Requirements for Commercial Food Waste Grinder Units	413.1
1010—98	Performance Requirements for Water Hammer Arresters	604.9
1011—95	Performance Requirements for Hose Connection Vacuum Breakers	Table 608.1, 608.13.6
1012—95	Performance Requirements for Backflow Preventers with Intermediate Atmospheric Vent	Table 608.1, 608.13.3, 608.16.2
1013—99	Performance Requirements for Reduced Pressure Principle Backflow Preventers and Reduced Pressure Fire Protection Principle Backflow Preventers	Table 608.1, 608.13.2, 608.16.2
1014—90	Performance Requirements for Handheld Showers	424.2
1015—99	Performance Requirements for Double Check Backflow Prevention Assemblies And Double Check Fire Protection Backflow Prevention Assemblies	Table 608.1, 608.13.7
1016—96	Performance Requirements for Individual Thermostatic, Pressure Balancing and Combination Control Valves for Bathing Facilities	424.3

ASSE—continued

1017—99	Performance Requirements for Temperature Actuated Mixing Valves for Hot Water Distribution Systems	424.3, 424.5, 613.1
1018—86	Performance Requirements for Trap Seal Primer Valves; Water Supply Fed	1002.4
1019—97	Performance Requirements for Wall Hydrants, Freezeless, Automatic Draining, Anti-Backflow Types	Table 608.1, 608.13.6
1020—98	Performance Requirements for Pressure Vacuum Breaker Assembly	Table 608.1, 608.13.5
1022—98	Performance Requirements for Backflow Preventer for Carbonated Beverage Machines	Table 608.1, 608.16.1
1024—98	Performance Requirements for Dual Check Valve Type Backflow Preventers (for Residential Supply Service or Individual Outlets)	605.3.1, Table 608.1
1035—95	Performance Requirements for Laboratory Faucet Backflow Preventers	Table 608.1, 608.13.6
1037—90	Performance Requirements for Pressurized Flushing Devices for Plumbing Fixtures	425.2
1044—86	Performance Requirements for Trap Seal Primer Valves; Drainage Type	1002.4
1047—99	Performance Requirements for Reduced Pressure Detector Fire Protection Backflow Prevention Assemblies	Table 608.1, 608.13.2
1048—99	Performance Requirements for Double Check Detector Fire Protection Backflow Prevention Assemblies	Table 608.1, 608.13.7
1051—98	Performance Requirements for Air Admittance Valves for Plumbing Drainage Systems-Fixture and Branch Devices	917.1
1052—94	Performance Requirements for Hose Connection Backflow Preventers	Table 608.1, 608.13.6
1055—98	Performance Requirements for Backflow Devices for Chemical Dispensing Systems	608.13.9
1056—95	Performance Requirements for Back Siphonage Vacuum Breaker	Table 608.1, 608.13.5, 608.13.8
1060—96	Performance Requirements for Outdoor Enclosures for Backflow Prevention Assemblies	608.14.1
1062—97	Performance Requirements for Temperature Actuated, Flow Reduction Valves to Individual Fixture Fittings	424.5
1066—97	Performance Requirements for Individual Pressure Balancing Valves for Individual Fixture Fittings	604.11
5013—98	Performance Requirements for Testing Reduced Pressure Principle Backflow Preventers (RP) and Reduced Pressure Fire Protection Principle Backflow Preventers (RFP)	312.9.2
5015—98	Performance Requirements for Testing Double Check Backflow Prevention Assemblies (DC) and Double Check Fire Protection Backflow Prevention Assemblies (DCF)	312.9.2
5020—98	Performance Requirements for Testing Pressure Vacuum Breaker Assembly (PVBA)	312.9.2
5047—98	Performance Requirements for Testing Reduced Pressure Detector Fire Protection Backflow Prevention Assemblies (RPDF)	312.9.2
5048—98	Performance Requirements for Testing Double Check Detector Fire Protection Backflow Prevention Assemblies (DCDF)	312.9.2
5052—98	Performance Requirements for Testing Hose Connection Backflow Preventers	312.9.2
5056—98	Performance Requirements for Testing Spill Resistant Vacuum Breaker	312.9.2

ASTM

ASTM International
100 Barr Harbor Drive
West Conshohocken, PA 19428-2959

Standard Reference Number	Title	Referenced in code section number
A 53/A 53M—01	Specification for Pipe, Steel, Black and Hot-Dipped, Zinc-Coated Welded and Seamless	Table 605.3, Table 605.4, Table 702.1
A 74—98	Specification for Cast Iron Soil Pipe and Fittings	Table 702.1, Table 702.2, Table 702.3, Table 702.4, 708.2, Table 1102.4, Table 1102.5, Table 1102.7
A 312/A 312M—01	Specification for Seamless and Welded Austenitic Stainless Steel Pipes	Table 605.4, Table 605.5, Table 605.6, 605.23.2
A 733—01	Specification for Welded and Seamless Carbon Steel and Austenitic Stainless Steel Pipe Nipples	Table 605.8
A 778—01	Specification for Welded Unannealed Austenitic Stainless Steel Tubular Products	Table 605.4, Table 605.5, Table 605.6
A 888—98e1	Specification for Hubless Cast Iron Soil Pipe and Fittings for Sanitary and Storm Drain, Waste, and Vent Piping Application	Table 702.1, Table 702.2, Table 702.3, Table 702.4, Table 1102.4, Table 1102.5, Table 1102.7
B 32—00	Specification for Solder Metal	605.14.3, 605.15.4, 705.9.3, 705.10.3
B 42—98	Specification for Seamless Copper Pipe, Standard Sizes	Table 605.3, Table 605.4, Table 702.1

ASTM—continued

D 2662—96a	Specification for Polybutylene (PB) Plastic Pipe (SDR-PR) Based on Controlled Inside Diameter	Table 605.3
D 2665—01	Specification for Poly (Vinyl Chloride) (PVC) Plastic Drain, Waste, and Vent Pipe and Fittings	Table 702.1, Table 702.2, Table 702.3, Table 702.4, Table 1102.4, Table 1102.7
D 2666—96a	Specification for Polybutylene (PB) Plastic Tubing	Table 605.3
D 2672—96a	Specification for Joints for IPS PVC Pipe Using Solvent Cement	Table 605.3
D 2729—96a	Specification for Poly (Vinyl Chloride) (PVC) Sewer Pipe and Fittings	Table 1102.5
D 2737—01	Specification for Polyethylene (PE) Plastic Tubing	Table 605.3
D 2751—96a	Specification for Acrylonitrile-Butadiene-Styrene (ABS) Sewer Pipe and Fittings	Table 702.3, Table 1102.4
D 2846/D 2846M—99	Specification for Chlorinated Poly (Vinyl Chloride) (CPVC) Plastic Hot and Cold Water Distribution Systems	Table 605.3, Table 605.4, 605.16.2
D 2855—96	Standard Practice for Making Solvent-Cemented Joints with Poly (Vinyl Chloride) (PVC) Pipe and Fittings	605.21.2, 705.8.2, 705.14.2
D 2949—00a	Specification for 3.25-In Outside Diameter Poly (Vinyl Chloride) (PVC) Plastic Drain, Waste, and Vent Pipe and Fittings	Table 702.1, Table 702.2, Table 702.3
D 3034—00	Specification for Type PSM Poly (Vinyl Chloride) (PVC) Sewer Pipe and Fittings	Table 702.3, Table 702.4, Table 1102.4
D 3139—98	Specification for Joints for Plastic Pressure Pipes Using Flexible Elastomeric Seals	605.10.1, 605.21.1
D 3212—96a	Specification for Joints for Drain and Sewer Plastic Pipes Using Flexible Elastomeric Seals	705.2.1, 705.7.1, 705.8.1, 705.14.1
D 3309—96a	Specification for Polybutylene (PB) Plastic Hot and Cold Water Distribution Systems	Table 605.3, Table 605.4, 605.19.2, 605.19.3
D 3311—94	Specification for Drain, Waste and Vent (DWV) Plastic Fittings Patterns	Table 702.4
D 4068—01	Specification for Chlorinated Polyethlene (CPE) Sheeting for Concealed Water-Containment Membrane	417.5.2.2
D 4551—96(2001)	Specification for Poly (Vinyl Chloride) (PVC) Plastic Flexible Concealed Water-Containment Membrane	417.5.2.1
F 405—97	Specification for Corrugated Polyethylene (PE) Tubing and Fittings	Table 1102.5
F 409—99a	Specification for Thermoplastic Accessible and Replaceable Plastic Tube and Tubular Fittings	424.1.2, Table 1102.7
F 437—99	Specification for Threaded Chlorinated Poly (Vinyl Chloride) (CPVC) Plastic Pipe Fittings, Schedule 80	Table 605.5, Table 1102.7
F 438—01	Specification for Socket-Type Chlorinated Poly (Vinyl Chloride) (CPVC) Plastic Pipe Fittings, Schedule 40	Table 605.5, Table 1102.7
F 439—01	Specification for Socket-Type Chlorinated Poly (Vinyl Chloride) (CPVC) Plastic Pipe Fittings, Schedule 80	Table 605.5, Table 1102.7
F 441/F 441M—99	Specification for Chlorinated Poly (Vinyl Chloride) (CPVC) Plastic Pipe, Schedules 40 and 80	Table 605.3, Table 605.4 Table 605.5
F 442/F 442M—99	Specification for Chlorinated Poly (Vinyl Chloride) (CPVC) Plastic Pipe (SDR-PR)	Table 605.3, Table 605.4, Table 605.5
F 477—99	Specification for Elastomeric Seals (Gaskets) for Joining Plastic Pipe	605.22, 705.16
F 493—97	Specification for Solvent Cements for Chlorinated Poly (Vinyl Chloride) (CPVC) Plastic Pipe and Fittings	605.16.2
F 628—01	Specification for Acrylonitrile-Butadiene-Styrene (ABS) Schedule 40 Plastic Drain, Waste, and Vent Pipe with a Cellular Core	Table 702.1, Table 702.2, Table 702.3, Table 702.4, 705.2.2, 705.7.2, Table 1102.4
F 656—96a	Specification for Primers for Use in Solvent Cement Joints of Poly (Vinyl Chloride) (PVC) Plastic Pipe and Fittings	605.21.2, 705.8.2, 705.14.2
F 714—00	Specification for Polyethylene (PE) Plastic Pipe (SDR-PR) Based on Outside Diameter	Table 702.3
F 876—00	Specification for Cross-linked Polyethylene (PEX) Tubing	Table 605.3
F 877—00	Specification for Cross-linked Polyethylene (PEX) Plastic Hot and Cold Water Distribution Systems	Table 605.3, Table 605.4
F 891—00	Specification for Coextruded Poly (Vinyl Chloride) (PVC) Plastic Pipe with a Cellular Core	Table 702.1, Table 702.2, Table 702.3, Table 702.4, Table 1102.4, Table 1102.5
F 1281—e01	Specification for Cross-Linked Polyethylene/Aluminum/Cross-Linked Polyethylene (PEX-AL-PEX) Pressure Pipe	Table 605.3, Table 605.4
F 1282—01a	Specification for Polyethylene/Aluminum/Polyethylene (PE-AL-PE) Composite Pressure Pipe	Table 605.3, Table 605.4
F 1488—00	Specification for Coextruded Composite Pipe	Table 702.1, Table 702.2, Table 702.3
F 1807—99	Specification for Metal Insert Fittings Utilizing a Copper Crimp Ring for SDR9 Cross-linked Polyethylene (PEX) Tubing	Table 605.5, 605.17.2
F 1866—98	Specification for Poly (Vinyl Chloride) (PVC) Plastic Schedule 40 Drainage and DWV Fabricated Fittings	Table 702.4

ASTM—continued

F 1960—99	Specification for Cold Expansion Fittings with PEX Reinforcing Rings for use with Cross-linked Polyethylene (PEX) Tubing.	Table 605.5
F 1974—00	Specification for Metal Insert Fittings for Polyethylene/Aluminum/Polyethylene and Cross-linked Polyethylene/Aluminum/Cross-linked Polyethylene Composite Pressure Pipe.	Table 605.5
F 2080—01	Specifications for Cold-Expansion Fittings with Metal Compression-Sleeves for Cross-linked Polyethylene (PEX) Pipe.	Table 605.5

AWS

American Welding Society
550 N.W. LeJeune Road
Miami, FL 33126

Standard Reference Number	Title	Referenced in code section number
A5.8—92	Specifications for Filler Metals for Brazing and Braze Welding	605.12.1, 605.14.1, 605.15.1, 705.4.1, 705.9.1, 705.10.1

AWWA

American Water Works Association
6666 West Quincy Avenue
Denver, CO 80235

Standard Reference Number	Title	Referenced in code section number
C104—95	Standard for Cement-Mortar Lining for Ductile-Iron Pipe and Fittings for Water	605.3, 605.5
C110—98	Standard for Ductile-Iron and Gray-Iron Fittings, 3 Inches through 48 Inches, for Water	Table 605.5, Table 702.4, Table 1102.7
C111—00	Standard for Rubber-Gasket Joints for Ductile-Iron Pressure Pipe and Fittings	605.13
C115—99	Standard for Flanged Ductile-Iron Pipe with Ductile-Iron or Gray-Iron Threaded Flanges	Table 605.3
C151—96	Standard for Ductile-Iron Pipe, Centrifugally Cast for Water	Table 605.3
C153—00	Standard for Ductile-Iron Compact Fittings for Water Service	Table 605.5
C510—00	Double Check Valve Backflow Prevention Assembly	Table 608.1, 608.13.7
C511—00	Reduced-Pressure Principle Backflow Prevention Assembly	Table 608.1, 608.13.2, 608.16.2
C651—99	Disinfecting Water Mains	610.1
C652—92	Disinfection of Water-Storage Facilities	610.1

CISPI

Cast Iron Soil Pipe Institute
5959 Shallowford Road, Suite 419
Chattanooga, TN 37421

Standard Reference Number	Title	Referenced in code section number
301—00	Specification for Hubless Cast Iron Soil Pipe and Fittings for Sanitary and Storm Drain, Waste and Vent Piping Applications	Table 702.1, Table 702.2, Table 702.3, Table 702.4, Table 1102.4, Table 1102.5, Table 1102.7
310—97	Specification for Coupling for Use in Connection with Hubless Cast Iron Soil Pipe and Fittings for Sanitary and Storm Drain, Waste and Vent Piping Applications	705.5.3

CSA

Canadian Standards Association
178 Rexdale Blvd.
Rexdale (Toronto), Ontario, Canada M9W 1R3

Standard Reference Number	Title	Referenced in code section number
B45.1—99	Ceramic Plumbing Fixtures	408.1, 416.1, 418.1, 419.1, 420.1
B45.2—99	Enameled Cast-Iron Plumbing Fixtures	407.1, 415.1, 416.1, 418.1
B45.3—99	Porcelain Enameled Steel Plumbing Fixtures	407.1, 416.1, 418.1

CSA—continued

B45.4—99	Stainless-Steel Plumbing Fixtures	415.1, 416.1, 418.1, 420.1
B45.5—99	Plastic Plumbing Fixtures	407.1, 416.2, 417.1, 419.1, 420.1, 421.1
B45.9—99	Macerating Systems and Related Components	712.4.1
B45.10—01	Hydromassage Bathtubs	421.1
B64.7—94	Vacuum Breakers, Laboratory Faucet Type (LFVB)	Table 608.1, 608.13.6
B79—94(2000)	Floor, Area and Shower Drains, and Cleanouts for Residential Construction	412.1
B125—98	Plumbing Fittings	424.1, 424.3, 424.4, 425.3.1, 425.4, Table 608.1
B137.1—99	Polyethylene Pipe, Tubing and Fittings for Cold Water Pressure Services	Table 605.3
B137.2—99	PVC Injection-Moulded Gasketed Fittings for Pressure Applications	Table 605.5, Table 1102.7
B137.3—99	Rigid Poly (Vinyl Chloride) (PVC) Pipe for Pressure Applications	Table 605.3, 605.21.2, 705.8.2, 705.14.2
B137.5—99	Cross-Linked Polyethylene (PEX) Tubing Systems for Pressure Applications—with Revisions through September 1992	Table 605.3, Table 605.4
B137.6—99	CPVC Pipe, Tubing and Fittings for Hot and Cold Water Distribution Systems—with Revisions through May 1986	Table 605.3, Table 605.4
B181.1—99	ABS Drain, Waste, and Vent Pipe and Pipe Fittings	Table 702.1, Table 702.2, Table 702.4, 705.2.2, 705.7.2, 715.2
B181.2—99	PVC Drain, Waste, and Vent Pipe and Pipe Fittings—with Revisions through December 1993	Table 702.1, Table 702.2, 705.8.2, 705.14.2, 715.2
B182.1—99	Plastic Drain and Sewer Pipe and Pipe Fittings	705.8.2, 705.14.2
B182.2—99	PVC Sewer Pipe and Fittings (PSM Type)	Table 702.3, Table 1102.4, Table 1102.5
CAN3-B137.8M—99	Polybutylene (PB) Piping for Pressure Applications—with Revisions through July 1992	Table 605.3, Table 605.4, 605.19.2, 605.19.3
CAN/CSA-A257.1M—92	Circular Concrete Culvert, Storm Drain, Sewer Pipe and Fittings	Table 702.3, Table 1102.4
CAN/CSA-A257.2M—92	Reinforced Circular Concrete Culvert, Storm Drain, Sewer Pipe and Fittings	Table 702.3, Table 1102.4
CAN/CSA-A257.3M—92	Joints for Circular Concrete Sewer and Culvert Pipe, Manhole Sections, and Fittings Using Rubber Gaskets	705.6, 705.16
CAN/CSA-B64.1.1—01	Vacuum Breakers, Atmospheric Type (AVB)	425.2, Table 608.1, 608.13.6
CAN/CSA-B64.2—01	Vacuum Breakers, Hose Connection Type (HCVB)	Table 608.1, 608.13.6
CAN/CSA-B64.2.2—01	Vacuum Breakers, Hose Connection Type (HCVB) with Automatic Draining Feature	Table 608.1, 608.13.6
CAN/CSA-B64.3—01	Backflow Preventers, Dual Check Valve Type with Atmospheric Port (DCAP)	Table 608.1, 608.13.3, 608.16.2
CAN/CSA-B64.4—01	Backflow Preventers, Reduced Pressure Principle Type (RP)	Table 608.1, 608.13.2, 608.16.2
CAN/CSA-B64.10—01	Manual for the Selection, Installation, Maintenance and Field Testing of Backflow Prevention Devices	312.9.2
CAN/CSA-B137.9—99	Polyethylene/Aluminum/Polyethylene Composite Pressure Pipe Systems	Table 605.3
CAN/CSA-B137.10M—99	Cross-linked Polyethylene/Aluminum/Polyethylene Composite Pressure Pipe Systems	Table 605.3, Table 605.4
CAN/CSA-B181.3—99	Polyolefin Laboratory Drainage Systems	Table 702.1, Table 702.2
CAN/CSA-B182.4—99	Profile PVC Sewer Pipe and Fittings	Table 702.3, Table 1102.4, Table 1102.5
CAN/CSA-B602—99	Mechanical Couplings for Drain, Waste, and Vent Pipe and Sewer Pipe	705.2.1, 705.5.3, 705.6, 705.7.1, 705.14.1, 705.15, 705.16

FS

Federal Specification
1941 Jefferson Davis Highway, Suite 104
Arlington, VA 22202

Standard Reference Number	Title	Referenced in code section number
TT-P-1536A(1975)	Federal Specification for Plumbing Fixture Setting Compound	405.4

* Standards are available from the Supt. of Documents, U.S. Government Printing Office, Washington, DC 20402-9325.

ICC

International Code Council
5203 Leesburg Pike, Suite 600
Falls Church, VA 22041

Standard Reference Number	Title	Referenced in code section number
IBC—03	International Building Code®	201.3, 305.4, 307.1, 307.2, 307.3, 308.2, 309.1, 310.1, 310.3, 403.1, Table 403.1, 404.1, 407.3, 417.6, 502.6, 606.5.2, 1106.5

ICC—continued

ICC EC—03	ICC Electrical Code™	201.3, 502.1, 504.3, 1113.1.3
IEBC—03	International Existing Building Code®	101.2
IECC—03	International Energy Conservation Code®	313.1, 607.2, 607.2.1
IFC—03	International Fire Code®	201.3, 1201.1
IFGC—03	International Fuel Gas Code®	101.2, 201.3, 502.1
IMC—03	International Mechanical Code®	201.3, 307.6, 310.1, 422.9, 502.1, 612.1, 1202.1
IPSDC—03	International Private Sewage Disposal Code®	701.2
IRC—03	International Residential Code®	101.2

ISEA

Industry Safety Equipment Association
1901 N. Moore Street, Suite 808
Arlington, VA 22209

Standard Reference Number	Title	Referenced in code section number
Z358.1—98	Emergency eyewash and shower equipment	411.1

NFPA

National Fire Protection Association
Batterymarch Park
Quincy, MA 02269

Standard Reference Number	Title	Referenced in code section number
50—01	Bulk Oxygen Systems at Consumer Sites	1203.1
51—97	Design and Installation of Oxygen-Fuel Gas Systems for Welding, Cutting, and Allied Processes	1203.1
70—99	National Electrical Code	502.1, 504.3, 1113.1.3
99C—99	Gas and Vacuum Systems	1202.1

NSF

National Sanitation Foundation
789 Dixboro Road
Ann Arbor, MI 48105

Standard Reference Number	Title	Referenced in code section number
3—1996	Commercial Spray-Type Dishwashing and Glasswashing Machines	409.1
18—1996a	Manual Food and Beverage Dispensing Equipment	426.1
14—1999	Plastic Piping System Components and Related Materials	303.3, 611.3
42—2000	Drinking Water Treatment Units—Aesthetic Effects	611.1, 611.3
44—2000	Residential Cation Exchange Water Softeners	611.1, 611.3
53—2001	Drinking Water Treatment Units—Health Effects	611.1, 611.3
58—2001	Reverse Osmosis Drinking Water Treatment Systems	611.2
61—2001	Drinking Water System Components—Health Effects	424.1, 605.3, 605.4, 605.5, 611.3
62—1999	Drinking Water Distillation Systems	611.1

PDI

Plumbing and Drainage Institute
45 Bristol Drive, Suite 101
South Easton, MA 02375

Standard Reference Number	Title	Referenced in code section number
G101(1998)	Testing and Rating Procedure for Grease Interceptors with Appendix of Sizing and Installation Data	1003.3.4

APPENDIX A
PLUMBING PERMIT FEE SCHEDULE

Permit Issuance

1. For issuing each permit . . . . $ _____

2. For issuing each supplemental permit. . . . . _____

Unit Fee Schedule

1. For each plumbing fixture or trap or set of fixtures on one trap (including water, drainage piping and backflow protection thereof) . . . . _____

2. For each building sewer and each trailer park sewer . . . . _____

3. Rainwater systems—per drain (inside building) . . . . _____

4. For each cesspool (where permitted) . . . . _____

5. For each private sewage disposal system . . . . _____

6. For each water heater and/or vent . . . . _____

7. For each industrial waste pretreatment interceptor including its trap and vent, excepting kitchen-type grease interceptors functioning as fixture traps . . . . _____

8. For installation, alteration or repair of water-piping and/or water-treating equipment, each . . . . . _____

9. For repair or alteration of drainage or vent piping, each fixture . . . . _____

10. For each lawn sprinkler system on any one meter including backflow protection devices therefor . . . . _____

11. For atmospheric-type vacuum breakers not included in Item 2:

    1 to 5 . . . . _____

    over 5, each . . . . _____

12. For each backflow protective device other than atmospheric-type vacuum breakers:

    2 inches (51 mm) and smaller . . . . _____

    Over 2 inches (51 mm) . . . . _____

Other Inspections and Fees

1. Inspections outside of normal business hours . . . . _____ per hour (minimum charge two hours)

2. Reinspection fee assessed under provisions of Section 107.3.3 . . . . _____ each

3. Inspections for which no fee is specifically indicated . . . . _____ per hour (minimum charge one-half hour)

4. Additional plan review required by changes, additions or revisions to approved plans (minimum charge one-half hour) . . . . _____ per hour

APPENDIX B
RATES OF RAINFALL FOR VARIOUS CITIES

Rainfall rates, in inches per hour, are based on a storm of one-hour duration and a 100-year return period. The rainfall rates shown in the appendix are derived from Figure 1106.1.

Alabama:
Birmingham 3.8
Huntsville 3.6
Mobile 4.6
Montgomery 4.2

Alaska:
Fairbanks 1.0
Juneau 0.6

Arizona:
Flagstaff 2.4
Nogales. 3.1
Phoenix. 2.5
Yuma 1.6

Arkansas:
Fort Smith 3.6
Little Rock 3.7
Texarkana. 3.8

California:
Barstow. 1.4
Crescent City. 1.5
Fresno 1.1
Los Angeles 2.1
Needles. 1.6
Placerville 1.5
San Fernando. 2.3
San Francisco. 1.5
Yreka. 1.4

Colorado:
Craig 1.5
Denver 2.4
Durango 1.8
Grand Junction 1.7
Lamar 3.0
Pueblo 2.5

Connecticut:
Hartford 2.7
New Haven 2.8
Putnam 2.6

Delaware:
Georgetown 3.0
Wilmington. 3.1

District of Columbia:
Washington. 3.2

Florida:
Jacksonville 4.3
Key West 4.3
Miami 4.7
Pensacola. 4.6
Tampa 4.5

Georgia:
Atlanta 3.7
Dalton 3.4
Macon 3.9
Savannah 4.3
Thomasville 4.3

Hawaii:
Hilo. 6.2
Honolulu 3.0
Wailuku. 3.0

Idaho:
Boise 0.9
Lewiston 1.1
Pocatello 1.2

Illinois:
Cairo 3.3
Chicago. 3.0
Peoria. 3.3
Rockford 3.2
Springfield 3.3

Indiana:
Evansville 3.2
Fort Wayne 2.9
Indianapolis 3.1

Iowa:
Davenport 3.3
Des Moines. 3.4
Dubuque 3.3
Sioux City 3.6

Kansas:
Atwood. 3.3
Dodge City 3.3
Topeka 3.7
Wichita 3.7

Kentucky:
Ashland. 3.0
Lexington. 3.1
Louisville. 3.2
Middlesboro 3.2
Paducah. 3.3

Louisiana:
Alexandria. 4.2
Lake Providence . . . 4.0
New Orleans. 4.8
Shreveport 3.9

Maine:
Bangor 2.2
Houlton. 2.1
Portland. 2.4

Maryland:
Baltimore. 3.2
Hagerstown. 2.8
Oakland. 2.7
Salisbury 3.1

Massachusetts:
Boston 2.5
Pittsfield 2.8
Worcester. 2.7

Michigan:
Alpena 2.5
Detroit 2.7
Grand Rapids. 2.6
Lansing. 2.8
Marquette 2.4
Sault Ste. Marie . . . 2.2

Minnesota:
Duluth 2.8
Grand Marais. 2.3
Minneapolis 3.1
Moorhead. 3.2
Worthington 3.5

Mississippi:
Biloxi. 4.7
Columbus. 3.9
Corinth 3.6
Natchez. 4.4
Vicksburg. 4.1

Missouri:
Columbia. 3.2
Kansas City. 3.6
Springfield 3.4
St. Louis 3.2

Montana:
Ekalaka 2.5
Havre. 1.6
Helena 1.5
Kalispell 1.2
Missoula 1.3

Nebraska:
North Platte. 3.3
Omaha 3.8
Scottsbluff 3.1
Valentine 3.2

Nevada:
Elko. 1.0
Ely 1.1
Las Vegas. 1.4
Reno 1.1

New Hampshire:
Berlin. 2.5
Concord 2.5
Keene. 2.4

New Jersey:
Atlantic City 2.9
Newark 3.1
Trenton 3.1

New Mexico:
Albuquerque 2.0
Hobbs. 3.0
Raton 2.5
Roswell. 2.6
Silver City 1.9

New York:
Albany 2.5
Binghamton 2.3
Buffalo 2.3
Kingston 2.7
New York. 3.0
Rochester. 2.2

North Carolina:
Asheville 4.1
Charlotte 3.7
Greensboro 3.4
Wilmington. 4.2

North Dakota:
Bismarck 2.8
Devils Lake. 2.9
Fargo 3.1
Williston 2.6

Ohio:
Cincinnati 2.9
Cleveland. 2.6
Columbus. 2.8
Toledo 2.8

Oklahoma:
Altus 3.7
Boise City 3.3
Durant 3.8
Oklahoma City 3.8

Oregon:
Baker 0.9
Coos Bay 1.5
Eugene 1.3
Portland. 1.2

Pennsylvania:
Erie 2.6
Harrisburg 2.8
Philadelphia 3.1
Pittsburgh 2.6
Scranton 2.7

Rhode Island:
Block Island 2.75
Providence 2.6

South Carolina:
Charleston 4.3
Columbia 4.0
Greenville 4.1

South Dakota:
Buffalo 2.8
Huron 3.3
Pierre 3.1
Rapid City 2.9
Yankton 3.6

Tennessee:
Chattanooga 0.5
Knoxville 3.2
Memphis 3.7
Nashville 3.3

Texas:
Abilene 3.6
Amarillo 3.5
Brownsville 4.5
Dallas 4.0
Del Rio 4.0
El Paso 2.3
Houston 4.6

Lubbock 3.3
Odessa 3.2
Pecos 3.0
San Antonio 4.2

Utah:
Brigham City 1.2
Roosevelt 1.3
Salt Lake City 1.3
St. George 1.7

Vermont:
Barre 2.3
Bratteboro 2.7
Burlington 2.1
Rutland 2.5

Virginia:
Bristol 2.7
Charlottesville . . . 2.8
Lynchburg 3.2
Norfolk 3.4
Richmond 3.3

Washington:
Omak 1.1
Port Angeles 1.1
Seattle 1.4
Spokane 1.0
Yakima 1.1

West Virginia:
Charleston 2.8
Morgantown 2.7

Wisconsin:
Ashland 2.5
Eau Claire 2.9
Green Bay 2.6
La Crosse 3.1
Madison 3.0
Milwaukee 3.0

Wyoming:
Cheyenne 2.2
Fort Bridger 1.3
Lander 1.5
New Castle 2.5
Sheridan 1.7
Yellowstone Park . . . 1.4

APPENDIX C
GRAY WATER RECYCLING SYSTEMS

Note: Section 301.3 of this code requires all plumbing fixtures that receive water or waste to discharge to the sanitary drainage system of the structure. In order to allow for the utilization of a gray water recycling system, Section 301.3 should be revised to read as follows:

301.3 Connections to drainage system. All plumbing fixtures, drains, appurtenances and appliances used to receive or discharge liquid wastes or sewage shall be directly connected to the drainage system of the building or premises, in accordance with the requirements of this code. This section shall not be construed to prevent indirect waste systems provided for in Chapter 8.

Exception: Bathtubs, showers, lavatories, clothes washers and laundry sinks shall not be required to discharge to the sanitary drainage system where such fixtures discharge to an approved gray water recycling system.

C101
GRAY WATER RECYCLING SYSTEMS

C101.1 General. Gray water recycling systems shall receive the waste discharge only of bathtubs, showers, lavatories, clothes washers and laundry sinks. Recycled gray water shall be utilized only for flushing water closets and urinals that are located in the same building as the gray water recycling system. Such systems shall comply with Sections C101.2 through C101.12.

Exception: Gray water systems shall be permitted to be used for irrigation when specific approval is given by the authority having jurisdiction. Such systems shall be designed as required by Section 105.

C101.2 Definition. The following term shall have the meaning shown herein.

GRAY WATER. Waste water discharged from lavatories, bathtubs, showers, clothes washers and laundry sinks.

C101.3 Installation. All drain, waste and vent piping associated with gray water recycling systems shall be installed in full compliance with this code.

C101.4 Reservoir. Gray water shall be collected in an approved reservoir constructed of durable, nonabsorbent and corrosion-resistant materials. The reservoir shall be a closed and gas-tight vessel. Access openings shall be provided to allow inspection and cleaning of the reservoir interior. The holding capacity of the reservoir shall be a minimum of twice the volume of water required to meet the daily flushing requirements of the fixtures supplied with gray water, but not less than 50 gallons (189 L). The reservoir shall be sized to limit the retention time of gray water to 72 hours maximum.

C101.5 Filtration. Gray water entering the reservoir shall pass through an approved filter such as a media, sand or diatomaceous earth filter.

C101.6 Disinfection. Gray water shall be disinfected by an approved method that employs one or more disinfectants such as chlorine, iodine or ozone.

C101.7 Makeup water. Potable water shall be supplied as a source of makeup water for the gray water system. The potable water supply shall be protected against backflow in accordance with Section 608. There shall be a full-open valve on the makeup water supply line to the reservoir.

C101.8 Overflow. The collection reservoir shall be equipped with an overflow pipe of the same diameter as the influent pipe for the gray water. The overflow shall be directly connected to the sanitary drainage system.

C101.9 Drain. A drain shall be located at the lowest point of the collection reservoir and shall be directly connected to the sanitary drainage system. The drain shall be the same diameter as the overflow pipe required by Section C101.8 and shall be provided with a full-open valve.

C101.10 Vent required. The reservoir shall be provided with a vent sized in accordance with Chapter 9 based on the size of the reservoir influent pipe.

C101.11 Coloring. The gray water shall be dyed blue or green with a food grade vegetable dye before such water is supplied to the fixtures.

C101.12 Identification. All gray water distribution piping and reservoirs shall be identified as containing nonpotable water. Piping identification shall be in accordance with Section 608.8.

APPENDIX D
DEGREE DAY AND DESIGN TEMPERATURES

TABLE D101
DEGREE DAY AND DESIGN TEMPERATURES[a] FOR CITIES IN THE UNITED STATES

STATE	STATION[b]	HEATING DEGREE DAYS (yearly total)	Winter 97½%	Summer Dry bulb 2½%	Summer Wet bulb 2½%	DEGREES NORTH LATTITUDE[c]
AL	Birmingham	2,551	21	94	77	33°30'
	Huntsville	3,070	16	96	77	34°40'
	Mobile	1,560	29	93	79	30°40'
	Montgomery	2,291	25	95	79	32°20'
AK	Anchorage	10,864	-18	68	59	61°10'
	Fairbanks	14,279	-47	78	62	64°50'
	Juneau	9,075	1	70	59	58°20'
	Nome	14,171	-27	62	56	64°30'
AZ	Flagstaff	7,152	4	82	60	35°10'
	Phoenix	1,765	34	107	75	33°30'
	Tuscon	1,800	32	102	71	33°10'
	Yuma	974	39	109	78	32°40'
AR	Fort Smith	3,292	17	98	79	35°20'
	Little Rock	3,219	20	96	79	34°40'
	Texarkana	2,533	23	96	79	33°30'
CA	Fresno	2,611	30	100	71	36°50'
	Long Beach	1,803	43	80	69	33°50'
	Los Angeles	2,061	43	80	69	34°00'
	Los Angeles[d]	1,349	40	89	71	34°00'
	Oakland	2,870	36	80	64	37°40'
	Sacramento	2,502	32	98	71	38°30'
	San Diego	1,458	44	80	70	32°40'
	San Francisco	3,015	38	77	64	37°40'
	San Francisco[d]	3,001	40	71	62	37°50'
CO	Alamosa	8,529	-16	82	61	37°30'
	Colorado Springs	6,423	2	88	62	38°50'
	Denver	6,283	1	91	63	39°50'
	Grand Junction	5,641	7	94	63	39°10'
	Pueblo	5,462	0	95	66	38°20'
CT	Bridgeport	5,617	9	84	74	41°10'
	Hartford	6,235	7	88	75	41°50'
	New Haven	5,897	7	84	75	41°20'
DE	Wilmington	4,930	14	89	76	39°40'
DC	Washington	4,224	17	91	77	38°50'
FL	Daytona	879	35	90	79	29°10'
	Fort Myers	442	44	92	79	26°40'
	Jacksonville	1,239	32	94	79	30°30'
	Key West	108	57	90	79	24°30'
	Miami	214	47	90	79	25°50'
	Orlando	766	38	93	78	28°30'
	Pensacola	1,463	29	93	79	30°30'
	Tallahassee	1,485	30	92	78	30°20'
	Tampa	683	40	91	79	28°00'
	West Palm Beach	253	45	91	79	26°40'
GA	Athens	2,929	22	92	77	34°00'
	Atlanta	2,961	22	92	76	33°40'
	Augusta	2,397	23	95	79	33°20'
	Columbus	2.383	24	93	78	32°30'
	Macon	2,136	25	93	78	32°40'
	Rome	3,326	22	93	78	34°20'
	Savannah	1,819	27	93	79	32°10'
HI	Hilo	0	62	83	74	19°40'
	Honolulu	0	63	86	75	21°20'

(Continued)

TABLE D101—continued
DEGREE DAY AND DESIGN TEMPERATURES[a] FOR CITIES IN THE UNITED STATES

STATE	STATION[b]	HEATING DEGREE DAYS (yearly total)	DESIGN TEMPERATURES			DEGREES NORTH LATTITUDE[c]
			Winter	Summer		
			97½%	Dry bulb 2½%	Wet bulb 2½%	
ID	Boise	5,809	10	94	66	43°30′
	Lewiston	5,542	6	93	66	46°20′
	Pocatello	7,033	-1	91	63	43°00′
IL	Chicago (Midway)	6,155	0	91	75	41°50′
	Chicago (O'Hare)	6,639	-4	89	76	42°00′
	Chicago[d]	5,882	2	91	77	41°50′
	Moline	6,408	-4	91	77	41°30′
	Peoria	6,025	-4	89	76	40°40′
	Rockford	6,830	-4	89	76	42°10′
	Springfield	5,429	2	92	77	39°50′
IN	Evansville	4,435	9	93	78	38°00′
	Fort Wayne	6,205	1	89	75	41°00′
	Indianapolis	5,699	2	90	76	39°40′
	South Bend	6,439	1	89	75	41°40′
IA	Burlington	6,114	-3	91	77	40°50′
	Des Moines	6,588	-5	91	77	41°30′
	Dubuque	7,376	-7	88	75	42°20′
	Sioux City	6,951	-7	92	77	42°20′
	Waterloo	7,320	-10	89	77	42°30′
KS	Dodge City	4,986	5	97	73	37°50′
	Goodland	6,141	0	96	70	39°20′
	Topeka	5,182	4	96	78	39°00′
	Wichita	4,620	7	98	76	37°40′
KY	Covington	5,265	6	90	75	39°00′
	Lexington	4,683	8	91	76	38°00′
	Louisville	4,660	10	93	77	38°10′
LA	Alexandria	1,921	27	94	79	31°20′
	Baton Rouge	1,560	29	93	80	30°30′
	Lake Charles	1,459	31	93	79	30°10′
	New Orleans	1,385	33	92	80	30°00′
	Shreveport	2,184	25	96	79	32°30′
ME	Caribou	9,767	-13	81	69	46°50′
	Portland	7,511	-1	84	72	43°40′
MD	Baltimore	4,654	13	91	77	39°10′
	Baltimore[d]	4,111	17	89	78	39°20′
	Frederick	5,087	12	91	77	39°20′
MA	Boston	5,634	9	88	74	42°20′
	Pittsfield	7,578	-3	84	72	42°30′
	Worcester	6,969	4	84	72	42°20′
MI	Alpena	8,506	-6	85	72	45°00′
	Detroit (City)	6,232	6	88	74	42°20′
	Escanaba[d]	8,481	-7	83	71	45°40′
	Flint	7,377	1	87	74	43°00′
	Grand Rapids	6,894	5	88	74	42°50′
	Lansing	6,909	1	87	74	42°50′
	Marquette[d]	8,393	-8	81	70	46°30′
	Muskegon	6,696	6	84	73	43°10′
	Sault Ste. Marie	9,048	-8	81	70	46°30′
MN	Duluth	10,000	-16	82	70	46°50′
	Minneapolis	8,382	-12	89	75	44°50′
	Rochester	8,295	-12	87	75	44°00′
MS	Jackson	2,239	25	95	78	32°20′
	Meridian	2,289	23	95	79	32°20′
	Vicksburg[d]	2,041	26	95	80	32°20′

(Continued)

TABLE D101—continued
DEGREE DAY AND DESIGN TEMPERATURES[a] FOR CITIES IN THE UNITED STATES

STATE	STATION[b]	HEATING DEGREE DAYS (yearly total)	Winter 97$^1/_2$%	Summer Dry bulb 2$^1/_2$%	Summer Wet bulb 2$^1/_2$%	DEGREES NORTH LATTITUDE[c]
MO	Columbia	5,046	4	94	77	39°00′
	Kansas City	4,711	6	96	77	39°10′
	St. Joseph	5,484	2	93	79	39°50′
	St. Louis	4,900	6	94	77	38°50′
	St. Louis[d]	4,484	8	94	77	38°40′
	Springfield	4,900	9	93	77	37°10′
MT	Billings	7,049	-10	91	66	45°50′
	Great Falls	7,750	-15	88	62	47°30′
	Helena	8,129	-16	88	62	46°40′
	Missoula	8,125	-6	88	63	46°50′
NE	Grand Island	6,530	-3	94	74	41°00′
	Lincoln[d]	5,864	-2	95	77	40°50′
	Norfolk	6,979	-4	93	77	42°00′
	North Platte	6,684	-4	94	72	41°10′
	Omaha	6,612	-3	91	77	41°20′
	Scottsbluff	6,673	-3	92	68	41°50′
NV	Elko	7,433	-2	92	62	40°50′
	Ely	7,733	-4	87	59	39°10′
	Las Vegas	2,709	28	106	70	36°10′
	Reno	6,332	10	92	62	39°30′
	Winnemucca	6,761	3	94	62	40°50′
NH	Concord	7,383	-3	87	73	43°10′
NJ	Atlantic City	4,812	13	89	77	39°30′
	Newark	4,589	14	91	76	40°40′
	Trenton[d]	4,980	14	88	76	40°10′
NM	Albuquerque	4,348	16	94	65	35°00′
	Raton	6,228	1	89	64	36°50′
	Roswell	3,793	18	98	70	33°20′
	Silver City	3,705	10	94	64	32°40′
NY	Albany	6,875	-1	88	74	42°50′
	Albany[d]	6,201	1	88	74	42°50′
	Binghamton	7,286	1	83	72	42°10′
	Buffalo	7,062	6	85	73	43°00′
	NY (Cent. Park)[d]	4,871	15	89	75	40°50′
	NY (Kennedy)	5,219	15	87	75	40°40′
	NY (LaGuardia)	4,811	15	89	75	40°50′
	Rochester	6,748	5	88	73	43°10′
	Schenectady[d]	6,650	1	87	74	42°50′
	Syracuse	6,756	2	87	73	43°10′
NC	Charlotte	3,181	22	93	76	35°10′
	Greensboro	3,805	18	91	76	36°10′
	Raleigh	3,393	20	92	77	35°50′
	Winston-Salem	3,595	20	91	75	36°10′
ND	Bismarck	8,851	-19	91	71	46°50′
	Devils Lake[d]	9,901	-21	88	71	48°10′
	Fargo	9,226	-18	89	74	46°50′
	Williston	9,243	-21	88	70	48°10′
OH	Akron-Canton	6,037	6	86	73	41°00′
	Cincinnati[d]	4,410	6	90	75	39°10′
	Cleveland	6,351	5	88	74	41°20′
	Columbus	5,660	5	90	75	40°00′
	Dayton	5,622	4	89	75	39°50′
	Mansfield	6,403	5	87	74	40°50′
	Sandusky[d]	5,796	6	91	74	41°30′
	Toledo	6,494	1	88	75	41°40′
	Youngstown	6,417	4	86	73	41°20′

(Continued)

TABLE D101—continued
DEGREE DAY AND DESIGN TEMPERATURES[a] FOR CITIES IN THE UNITED STATES

STATE	STATION[b]	HEATING DEGREE DAYS (yearly total)	DESIGN TEMPERATURES				DEGREES NORTH LATTITUDE[c]
			Winter	Summer			
			97½%	Dry bulb 2½%	Wet bulb 2½%		
OK	Oklahoma City	3,725	13	97	77	35°20'
	Tulsa	3,860	13	98	78	36°10'
OR	Eugene	4,726	22	89	67	44°10'
	Medford	5,008	23	94	68	42°20'
	Portland	4,635	23	85	67	45°40'
	Portland[d]	4,109	24	86	67	45°30'
	Salem	4,754	23	88	68	45°00'
PA	Allentown	5,810	9	88	75	40°40'
	Erie	6,451	9	85	74	42°10'
	Harrisburg	5,251	11	91	76	40°10'
	Philadelphia	5,144	14	90	76	39°50'
	Pittsburgh	5,987	5	86	73	40°30'
	Pittsburgh[d]	5,053	7	88	73	40°30'
	Reading[d]	4,945	13	89	75	40°20'
	Scranton	6,254	5	87	73	41°20'
	Williamsport	5,934	7	89	74	41°10'
RI	Providence	5,954	9	86	74	41°40'
SC	Charleston	2,033	27	91	80	32°50'
	Charleston[d]	1,794	28	92	80	32°50'
	Columbia	2,484	24	95	78	34°00'
SD	Huron	8,223	-14	93	75	44°30'
	Rapid City	7,345	-7	92	69	44°00'
	Sioux Falls	7,839	-11	91	75	43°40'
TN	Bristol	4,143	14	89	75	36°30'
	Chattanooga	3,254	18	93	77	35°00'
	Knoxville	3,494	19	92	76	35°50'
	Memphis	3,232	18	95	79	35°00'
	Nashville	3,578	14	94	77	36°10'
TX	Abilene	2,624	20	99	74	32°30'
	Austin	1,711	28	98	77	30°20'
	Dallas	2,363	22	100	78	32°50'
	El Paso	2,700	24	98	68	31°50'
	Houston	1,396	32	94	79	29°40'
	Midland	2,591	21	98	72	32°00'
	San Angelo	2,255	22	99	74	31°20'
	San Antonio	1,546	30	97	76	29°30'
	Waco	2,030	26	99	78	31°40'
	Wichita Falls	2,832	18	101	76	34°00'
UT	Salt Lake City	6,052	8	95	65	40°50'
VT	Burlington	8,269	-7	85	72	44°30'
VA	Lynchburg	4,166	16	90	76	37°20'
	Norfolk	3,421	22	91	78	36°50'
	Richmond	3,865	17	92	78	37°30'
	Roanoke	4,150	16	91	74	37°20'
WA	Olympia	5,236	22	83	66	47°00'
	Seattle-Tacoma	5,145	26	80	64	47°30'
	Seattle[d]	4,424	27	82	67	47°40'
	Spokane	6,655	2	90	64	47°40'
WV	Charleston	4,476	11	90	75	38°20'
	Elkins	5,675	6	84	72	38°50'
	Huntington	4,446	10	91	77	38°20'
	Parkersburg[d]	4,754	11	90	76	39°20'

(Continued)

TABLE D101—continued
DEGREE DAY AND DESIGN TEMPERATURES[a] FOR CITIES IN THE UNITED STATES

| STATE | STATION[b] | HEATING DEGREE DAYS (yearly total) | DESIGN TEMPERATURES | | | DEGREES NORTH LATTITUDE[c] |
| | | | Winter | Summer | | |
			97$\frac{1}{2}$%	Dry bulb 2$\frac{1}{2}$%	Wet bulb 2$\frac{1}{2}$%	
WI	Green Bay	8,029	-9	85	74	44°30'
	La Crosse	7,589	-9	88	75	43°50'
	Madison	7,863	-7	88	75	43°10'
	Milwaukee	7,635	-4	87	74	43°00'
WY	Casper	7,410	-5	90	61	42°50'
	Cheyenne	7,381	-1	86	62	41°10'
	Lander	7,870	-11	88	63	42°50'
	Sheridan	7,680	-8	91	65	44°50'

a. All data was extracted from the 1985 ASHRAE Handbook, Fundamentals Volume.

b. Design data developed from airport temperature observations unless noted.

c. Latitude is given to the nearest 10 minutes. For example, the latitude for Miami, Florida, is given as 25°50' which is 25 degrees 50 minutes.

d. Design data developed from office locations within an urban area, not from airport temperature observations.

APPENDIX E
SIZING OF WATER PIPING SYSTEM

SECTION E101
GENERAL

E101.1 Scope.

E101.1.1 This appendix outlines two procedures for sizing a water piping system (see Sections E103.3 and E201.1). The design procedures are based on the minimum static pressure available from the supply source, the head charges in the system caused by friction and elevation, and the rates of flow necessary for operation of various fixtures.

E101.1.2 Because of the variable conditions encountered in hydraulic design, it is impractical to specify definite and detailed rules for sizing of the water piping system. Accordingly, other sizing or design methods conforming to good engineering practice standards are acceptable alternatives to those presented herein.

SECTION E102
INFORMATION REQUIRED

E102.1 Preliminary. Obtain the necessary information regarding the minimum daily static service pressure in the area where the building is to be located. If the building supply is to be metered, obtain information regarding friction loss relative to the rate of flow for meters in the range of sizes likely to be used. Friction loss data can be obtained from most manufacturers of water meters.

E102.2 Demand load.

E102.2.1 Estimate the supply demand of the building main and the principal branches and risers of the system by totaling the corresponding demand from the applicable part of Table E103.3(3).

E102.2.2 Estimate continuous supply demands in gallons per minute (L/m) for lawn sprinklers, air conditioners, etc., and add the sum to the total demand for fixtures. The result is the estimated supply demand for the building supply.

SECTION E103
SELECTION OF PIPE SIZE

E103.1 General. Decide from Table 604.3 what is the desirable minimum residual pressure that should be maintained at the highest fixture in the supply system. If the highest group of fixtures contains flush valves, the pressure for the group should not be less than 15 pounds per square inch (psi) (103.4 kPa) flowing. For flush tank supplies, the available pressure should not be less than 8 psi (55.2 kPa) flowing, except blowout action fixtures must not be less than 25 psi (172.4 kPa) flowing.

E103.2 Pipe sizing.

E103.2.1 Pipe sizes can be selected according to the following procedure or by other design methods conforming to acceptable engineering practice and approved by the administrative authority. The sizes selected must not be less than the minimum required by this code.

E103.2.2 Water pipe sizing procedures are based on a system of pressure requirements and losses, the sum of which must not exceed the minimum pressure available at the supply source. These pressures are as follows:

1. Pressure required at fixture to produce required flow. See Section 604.3 and Section 604.5.

2. Static pressure loss or gain (due to head) is computed at 0.433 psi per foot (9.8 kPa/m) of elevation change.

   Example: Assume that the highest fixture supply outlet is 20 feet (6096 mm) above or below the supply source. This produces a static pressure differential of 20 feet by 0.433 psi/foot (2096 mm by 9.8 kPa/m) and an 8.66 psi (59.8 kPa) loss.

3. Loss through water meter. The friction or pressure loss can be obtained from meter manufacturers.

4. Loss through taps in water main.

5. Losses through special devices such as filters, softeners, backflow prevention devices and pressure regulators. These values must be obtained from the manufacturers.

6. Loss through valves and fittings. Losses for these items are calculated by converting to equivalent length of piping and adding to the total pipe length.

7. Loss due to pipe friction can be calculated when the pipe size, the pipe length and the flow through the pipe are known. With these three items, the friction loss can be determined. For piping flow charts not included, use manufacturers' tables and velocity recommendations.

   Note: For the purposes of all examples, the following metric conversions are applicable:

   1 cubic foot per minute = 0.4719 L/s

   1 square foot = 0.0929 m²

   1 degree = 0.0175 rad

   1 pound per square inch = 6.895 kPa

   1 inch = 25.4 mm

   1 foot = 304.8 mm

   1 gallon per minute = 3.785 L/m

E103.3 Segmented loss method. The size of water service mains, branch mains and risers by the segmented loss method, must be determined according to water supply demand [gpm (L/m)], available water pressure [psi (kPa)] and friction loss caused by the water meter and developed length of pipe [feet (m)], including equivalent length of fittings. This design procedure is based on the following parameters:

• Calculates the friction loss through each length of the pipe.

- Based on a system of pressure losses, the sum of which must not exceed the minimum pressure available at the street main or other source of supply.

- Pipe sizing based on estimated peak demand, total pressure losses caused by difference in elevation, equipment, developed length and pressure required at most remote fixture, loss through taps in water main, losses through fittings, filters, backflow prevention devices, valves and pipe friction.

Because of the variable conditions encountered in hydraulic design, it is impractical to specify definite and detailed rules for sizing of the water piping system. Current sizing methods do not address the differences in the probability of use and flow characteristics of fixtures between types of occupancies. Creating an exact model of predicting the demand for a building is impossible and final studies assessing the impact of water conservation on demand are not yet complete. The following steps are necessary for the segmented loss method.

1. **Preliminary.** Obtain the necessary information regarding the minimum daily static service pressure in the area where the building is to be located. If the building supply is to be metered, obtain information regarding friction loss relative to the rate of flow for meters in the range of sizes to be used. Friction loss data can be obtained from manufacturers of water meters. It is essential that enough pressure be available to overcome all system losses caused by friction and elevation so that plumbing fixtures operate properly. Section 604.6 requires the water distribution system to be designed for the minimum pressure available taking into consideration pressure fluctuations. The lowest pressure must be selected to guarantee a continuous, adequate supply of water. The lowest pressure in the public main usually occurs in the summer because of lawn sprinkling and supplying water for air-conditioning cooling towers. Future demands placed on the public main as a result of large growth or expansion should also be considered. The available pressure will decrease as additional loads are placed on the public system.

2. **Demand load.** Estimate the supply demand of the building main and the principal branches and risers of the system by totaling the corresponding demand from the applicable part of Table E103.3(3). When estimating peak demand sizing methods typically use water supply fixture units (w.s.f.u.)(see Table E103.3(2)). This numerical factor measures the load-producing effect of a single plumbing fixture of a given kind. The use of such fixture units can be applied to a single basic probability curve (or table), found in the various sizing methods (Table E103.3(3)). The fixture units are then converted into gallons per minute (L/m) flow rate for estimating demand.

   2.1. Estimate continuous supply demand in gallons per minute (L/m) for lawn sprinklers, air conditioners, etc., and add the sum to the total demand for fixtures. The result is the estimated supply demand for the building supply. Fixture units cannot be applied to constant use fixtures such as hose bibbs, lawn sprinklers and air conditioners. These types of fixtures must be assigned the gallon per minute (L/m) value.

3. **Selection of pipe size.** This water pipe sizing procedure is based on a system of pressure requirements and losses, the sum of which must not exceed the minimum pressure available at the supply source. These pressures are as follows:

   3.1. Pressure required at the fixture to produce required flow. See Section 604.3 and Section 604.5.

   3.2. Static pressure loss or gain (because of head) is computed at 0.433 psi per foot (9.8 kPa/m) of elevation change.

   3.3. Loss through a water meter. The friction or pressure loss can be obtained from the manufacturer.

   3.4. Loss through taps in water main [see Table E103.3(4)].

   3.5. Losses through special devices such as filters, softeners, backflow prevention devices and pressure regulators. These values must be obtained from the manufacturers.

   3.6. Loss through valves and fittings [see Tables E103.3(5) and E103.3(6)]. Losses for these items are calculated by converting to equivalent length of piping and adding to the total pipe length.

   3.7. Loss caused by pipe friction can be calculated when the pipe size, the pipe length and the flow through the pipe are known. With these three items, the friction loss can be determined using Figures E103.3(2) through E103.3(7). When using charts, use pipe inside diameters. For piping flow charts not included, use manufacturers' tables and velocity recommendations. Before attempting to size any water supply system, it is necessary to gather preliminary information which includes available pressure, piping material, select design velocity, elevation differences and developed length to most remote fixture. The water supply system is divided into sections at major changes in elevation or where branches lead to fixture groups. The peak demand must be determined in each part of the hot and cold water supply system which includes the corresponding water supply fixture unit and conversion to gallons per minute (L/m) flow rate to be expected through each section. Sizing methods require the determination of the "most hydraulically remote" fixture to compute the pressure loss caused by pipe and fittings. The hydraulically remote fixture represents the most downstream fixture along the circuit of piping requiring the most available pressure to operate properly. Consideration must be given to all pressure demands and losses, such as friction caused by pipe, fittings and equipment, elevation and the residual pressure required by Table 604.3. The two most common and frequent complaints about the water supply system operation are lack of adequate pressure and noise.

Problem: What size Type L copper water pipe, service and distribution will be required to serve a two-story factory building having on each floor, back-to-back, two toilet rooms each equipped with hot and cold water? The highest fixture is 21 feet (6401 mm) above the street main, which is tapped with a 2-inch (51 mm) corporation cock at which point the minimum pressure is 55 psi (379.2 kPa). In the building basement, a 2-inch (51 mm) meter with a maximum pressure drop of 11 psi (75.8 kPa) and 3-inch (76 mm) reduced pressure principle backflow preventer with a maximum pressure drop of 9 psi (621 kPa) are to be installed. The system is shown by Figure E103.3(1). To be determined are the pipe sizes for the service main and the cold and hot water distribution pipes.

Solution: A tabular arrangement such as shown in Table E103.3(1) should first be constructed. The steps to be followed are indicated by the tabular arrangement itself as they are in sequence, columns 1 through 10 and lines A through L.

Step 1

Columns 1 and 2: Divide the system into sections breaking at major changes in elevation or where branches lead to fixture groups. After point B [see Figure E103.3(1)], separate consideration will be given to the hot and cold water piping. Enter the sections to be considered in the service and cold water piping in Column 1 of the tabular arrangement. Column 1 of Table E103.3(1) provides a line-by-line recommended tabular arrangement for use in solving pipe sizing.

The objective in designing the water supply system is to ensure an adequate water supply and pressure to all fixtures and equipment. Column 2 provides the pounds per square inch (psi) to be considered separately from the minimum pressure available at the main. Losses to take into consideration are the following: the differences in elevations between the water supply source and the highest water supply outlet, meter pressure losses, the tap in main loss, special fixture devices such as water softeners and prevention devices and the pressure required at the most remote fixture outlet. The difference in elevation can result in an increase or decrease in available pressure at the main. Where the water supply outlet is located above the source, this results in a loss in the available pressure and is subtracted from the pressure at the water source. Where the highest water supply outlet is located below the water supply source, there will be an increase in pressure that is added to the available pressure of the water source.

Column 3: According to Table E103.3(3), determine the gpm (L/m) of flow to be expected in each section of the system. These flows range from 28.6 to 108 gpm. Load values for fixtures must be determined as water supply fixture units and then converted to a gallon-per-minute (gpm) rating to determine peak demand. When calculating peak demands, the water supply fixture units are added and then converted to the gallon-per-minute rating. For continuous flow fixtures such as hose bibbs and lawn sprinkler systems, add the gallon-per-minute demand to the intermittent demand of fixtures. For example, a total of 120 water supply fixture units is converted to a demand of 48 gallons per minute. Two hose bibbs × 5 gpm demand = 10 gpm. Total gpm rating = 48.0 gpm + 10 gpm = 58.0 gpm demand.

Step 2

Line A: Enter the minimum pressure available at the main source of supply in Column 2. This is 55 psi (379.2 kPa). The local water authorities generally keep records of pressures at different times of day and year. The available pressure can also be checked from nearby buildings or from fire department hydrant checks.

Line B: Determine from Section 604.3 the highest pressure required for the fixtures on the system, which is 15 psi (103.4 kPa), to operate a flushometer valve. The most remote fixture outlet is necessary to compute the pressure loss caused by pipe and fittings, and represents the most downstream fixture along the circuit of piping requiring the available pressure to operate properly as indicated by Table 604.3.

Line C: Determine the pressure loss for the meter size given or assumed. The total water flow from the main through the service as determined in Step 1 will serve to aid in the meter selected. There are three common types of water meters; the pressure losses are determined by the American Water Works Association Standards for displacement type, compound type and turbine type. The maximum pressure loss of such devices takes into consideration the meter size, safe operating capacity (gpm) and maximum rates for continuous operations (gpm). Typically, equipment imparts greater pressure losses than piping.

Line D: Select from Table E103.3(4) and enter the pressure loss for the tap size given or assumed. The loss of pressure through taps and tees in pounds per square inch (psi) are based on the total gallon-per-minute flow rate and size of the tap.

Line E: Determine the difference in elevation between the main and source of supply and the highest fixture on the system. Multiply this figure, expressed in feet, by 0.43 psi (2.9 kPa). Enter the resulting psi loss on Line E. The difference in elevation between the water supply source and the highest water supply outlet has a significant impact on the sizing of the water supply system. The difference in elevation usually results in a loss in the available pressure because the water supply outlet is generally located above the water supply source. The loss is caused by the pressure required to lift the water to the outlet. The pressure loss is subtracted from the pressure at the water source. Where the highest water supply outlet is located below the water source, there will be an increase in pressure which is added to the available pressure of the water source.

Lines F, G and H: The pressure losses through filters, backflow prevention devices or other special fixtures must be obtained from the manufacturer or estimated and entered on these lines. Equipment such as backflow prevention devices, check valves, water softeners, instantaneous or tankless water heaters, filters and strainers can impart a much greater pressure loss than the piping. The pressure losses can range from 8 psi to 30 psi.

Step 3

Line I: The sum of the pressure requirements and losses that affect the overall system (Lines B through H) is entered on this line. Summarizing the steps, all of the system losses are subtracted from the minimum water pressure. The remainder is the pressure available for friction, defined as the energy available to push the water through the pipes to each fixture. This force can be used as an average pressure loss, as long as the pressure available for friction is not exceeded. Saving a certain amount for available water supply pressures as an area incurs growth, or because of aging of the pipe or equipment added to the system is recommended.

Step 4

Line J: Subtract Line i from Line A. This gives the pressure that remains available from overcoming friction losses in the system. This figure is a guide to the pipe size that is chosen for each section, incorporating the total friction losses to the most remote outlet (measured length is called developed length).

Exception: When the main is above the highest fixture, the resulting psi must be considered a pressure gain (static head gain) and omitted from the sums of Lines B through H and added to Line J.

The maximum friction head loss that can be tolerated in the system during peak demand is the difference between the static pressure at the highest and most remote outlet at no-flow conditions and the minimum flow pressure required at that outlet. If the losses are within the required limits, then every run of pipe will also be within the required friction head loss. Static pressure loss is the most remote outlet in feet x 0.433 = loss in psi caused by elevation differences.

Step 5

Column 4: Enter the length of each section from the main to the most remote outlet (at Point E). Divide the water supply system into sections breaking at major changes in elevation or where branches lead to fixture groups.

Step 6

Column 5: When selecting a trial pipe size, the length from the water service or meter to the most remote fixture outlet must be measured to determine the developed length. However, in systems having a flush valve or temperature controlled shower at the top most floors the developed length would be from the water meter to the most remote flush valve on the system. A rule of thumb is that size will become progressively smaller as the system extends farther from the main source of supply. Trial pipe size may be arrived at by the following formula:

Line J (Pressure available to overcome pipe friction) × 100/equivalent length of run total developed length to most

remote fixture × percentage factor of 1.5 (note: a percentage factor is used only as an estimate for friction losses imposed for fittings for initial trial pipe size) = psi (average pressure drops per 100 feet of pipe).

For trial pipe size see Figure E 103.3(3) (Type L copper) based on 2.77 psi and a 108 gpm = $2\frac{1}{2}$ inches. To determine the equivalent length of run to the most remote outlet, the developed length is determined and added to the friction losses for fittings and valves. The developed lengths of the designated pipe sections are as follows:

A - B	54 ft
B - C	8 ft
C - D	13 ft
D - E	150 ft

Total developed length = 225 ft

The equivalent length of the friction loss in fittings and valves must be added to the developed length (most remote outlet). Where the size of fittings and valves is not known, the added friction loss should be approximated. A general rule that has been used is to add 50 percent of the developed length to allow for fittings and valves. For example, the equivalent length of run equals the developed length of run (225 ft × 1.5 = 338 ft). The total equivalent length of run for determining a trial pipe size is 338 feet.

Example: 9.36 (pressure available to overcome pipe friction) ×100/ 338 (Equivalent length of run = 225 × 1.5) = 2.77 psi (average pressure drop per 100 feet of pipe).

Step 7

Column 6: Select from Table E103.3(6) the equivalent lengths for the trial pipe size of fittings and valves on each pipe section. Enter the sum for each section in Column 6. (The number of fittings to be used in this example must be an estimate.) The equivalent length of piping is the developed length plus the equivalent lengths of pipe corresponding to friction head losses for fittings and valves. Where the size of fittings and valves is not known, the added friction head losses must be approximated. An estimate for this example is as follows:

COLD WATER PIPE SECTION	FITTINGS/VALVES	PRESSURE LOSS EXPRESSED AS EQUIVALENT LENGTH OF TUBE (FEET)	HOT WATER PIPE SECTION	FITTINGS/VALVES	PRESSURE LOSS EXPRESSED AS EQUIVALENT LENGTH OF TUBE (FEET)
A-B	3-2¹/₂″ Gate valves	3	A-B	3-2¹/₂″ Gate valves	3
	1-2¹/₂″ Side branch tee	12		1-2¹/₂″ Side branch tee	12
B-C	1-2¹/₂″ Straight run tee	0.5	B-C	1-2″ Straight run tee	7
				1-2″ 90-degree ell	0.5
C-F	1-2¹/₂″ Side branch tee	12	C-F	1-1¹/₂″ Side branch tee	7
C-D	1-2¹/₂″ 90-degree ell	7	C-D	1-1¹/₂″ 90-degree ell	4
D-E	1-2¹/₂″ Side branch tee	12	D-E	1-1¹/₂″ Side branch tee	7

Step 8

Column 7: Add the figures from Column 4 and Column 6, and enter in Column 7. Express the sum in hundreds of feet.

Step 9

Column 8: Select from Figure E103.3(3) the friction loss per 100 feet (30 480 mm) of pipe for the gallon-per-minute flow in a section (Column 3) and trial pipe size (Column 5). Maximum friction head loss per 100 feet is determined on the basis of total pressure available for friction head loss and the longest equivalent length of run. The selection is based on the gallon-per-minute demand, the uniform friction head loss, and the maximum design velocity. Where the size indicated by hydraulic table indicates a velocity in excess of the selected velocity, a size must be selected which produces the required velocity.

Step 10

Column 9: Multiply the figures in Columns 7 and 8 for each section and enter in Column 9.

Total friction loss is determined by multiplying the friction loss per 100 feet (30 480 mm) for each pipe section in the total developed length by the pressure loss in fittings expressed as equivalent length in feet. Note: Section C-F should be considered in the total pipe friction losses only if greater loss occurs in Section C-F than in pipe section D-E. Section C-F is not considered in the total developed length. Total friction loss in equivalent length is determined as follows:

Step 11

Line K: Enter the sum of the values in Column 9. The value is the total friction loss in equivalent length for each designated pipe section.

Step 12

Line L: Subtract Line J from Line K and enter in Column 10.

The result should always be a positive or plus figure. If it is not, repeat the operation using Columns 5, 6, 8 and 9 until a balance or near balance is obtained. If the difference between Lines J and K is a high positive number, it is an indication that the pipe sizes are too large and should be reduced, thus saving materials. In such a case, the operations using Columns 5, 6, 8 and 9 should again be repeated.

The total friction losses are determined and subtracted from the pressure available to overcome pipe friction for trial pipe size. This number is critical as it provides a guide to whether the pipe size selected is too large and the process should be repeated to obtain an economically designed system.

Answer: The final figures entered in Column 5 become the design pipe size for the respective sections. Repeating this operation a second time using the same sketch but considering the demand for hot water, it is possible to size the hot water distribution piping. This has been worked up as a part of the overall problem in the tabular arrangement used for sizing the service and water distribution piping. Note that consideration must be given to the pressure losses from the street main to the water heater (Section A-B) in determining the hot water pipe sizes.

PIPE SECTIONS	FRICTION LOSS EQUIVALENT LENGTH (feet)	
	Cold Water	Hot Water
A-B	0.69 x 3.2 = 2.21	0.69 x 3.2 = 2.21
B-C	0.085 x 3.1 = 0.26	0.16 x 1.4 = 0.22
C-D	0.20 x 1.9 = 0.38	0.17 x 3.2 = 0.54
D-E	1.62 x 1.9 = 3.08	1.57 x 3.2 = 5.02
Total pipe friction losses (Line K)	5.93	7.99

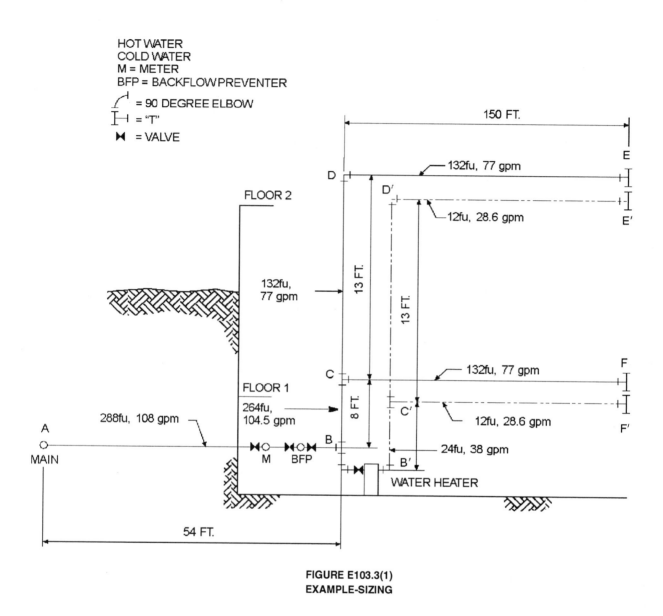

FIGURE E103.3(1)
EXAMPLE-SIZING

For SI: 1 foot = 304.8 mm, 1 gpm = 3.785 L/m.

TABLE E103.3(1)
RECOMMENDED TABULAR ARRANGEMENT FOR USE IN SOLVING PIPE SIZING PROBLEMS

COLUMN	1		2	3	4	5	6	7	8	9	10
Line	Description		Lb per square inch (psi)	Gal. per min through section	Length of section (feet)	Trial pipe size (inches)	Equivalent length of fittings and valves (feet)	Total equivalent length col. 4 and col. 6 (100 feet)	Friction loss per 100 feet of trial size pipe (psi))	Friction loss in equivalent length col. 8 x col. 7 (psi)	Excess pressure over friction losses (psi)
A	Service	Minimum pressure available at main........................55.00									
B	And	Highest pressure required at a fixture									
C	Cold	(Section 604.3)..15.00									
D	Water	Meter loss 2″ meter...11.00				....					
E	Distribution	Tap in main loss 2″ tap (Table E103A)......................1.61									
F	Piping[a]	Static head loss 21 x 43 psi......................................9.03									
G		Special fixture loss backflow preventer.....................9.00									
H		Special fixture loss—Filter.....................................0.00									
I		Special fixture loss—Other.....................................0.00									
		Total overall losses and requirements									
J		(Sum of Lines B through H....................................45.64									
		Pressure available to overcome pipe									
		Friction (Line A minus Lines B to H)........................9.36									
	DESIGNATION		FU								
	Pipe section (from Diagram)	AB.................................288	108.0	54	2½	15.00	0.69	3.2	2.21	—
	Cold water	BC.................................264	104.5	8	2½	0.5	0.85	3.1	0.26	—
	Distribution	CD.................................132	77.0	13	2½	7.00	0.20	1.9	0.38	—
	Piping	CF[b].............................132	77.0	150	2½	12.00	1.62	1.9	3.08	—
		DE[b].............................132	77.0	150	2½	12.00	1.62	1.9	3.08	—
K	Total pipe friction losses (cold)		—	—	—	—	—	—	—	5.93	—
L	Difference (Line J minus Line K)		—	—	—	—	—	—	—	—	3.43
	Pipe section (from diagram)	A′B′.................................288	108.0	54	2½	12.00	0.69	3.3	2.21	—
	Diagram	B′C′.................................24	38.0	8	2	7.5	0.16	1.4	0.22	—
	Hot Water	C′D′[b].............................12	28.6	13	1½	4.0	0.17	3.2	0.54	—
	Distribution	C′F′[b].............................12	28.6	150	1½	7.00	1.57	3.2	5.02	—
	Piping	D′E′[b].............................12	28.6	150	1½	7.00	1.57	3.2	5.02	—
K	Total pipe friction losses (hot)		—	—	—	—	—	—	—	7.99	—
L	Difference (line) Minus Line K		—	—	—	—	—	—	—	—	1.37

For SI:1 inch = 25.4 mm, 1 foot = 304.8 mm, 1 psi = 6.895 kPa, 1 gpm = 3.785 L/m.
a. To be considered as pressure gain for fixtures below main (to consider separately, omit from "I" and add to "J").
b. To consider separately, in K use C-F only if greater loss than above.

TABLE E103.3(2)
LOAD VALUES ASSIGNED TO FIXTURES[a]

FIXTURE	OCCUPANCY	TYPE OF SUPPLY CONTROL	LOAD VALUES, IN WATER SUPPLY FIXTURE UNITS (wsfu)		
			Cold	Hot	Total
Bathroom group	Private	Flush tank	2.7	1.5	3.6
Bathroom group	Private	Flush valve	6.0	3.0	8.0
Bathtub	Private	Faucet	1.0	1.0	1.4
Bathtub	Public	Faucet	3.0	3.0	4.0
Bidet	Private	Faucet	1.5	1.5	2.0
Combination fixture	Private	Faucet	2.25	2.25	3.0
Dishwashing machine	Private	Automatic	—	1.4	1.4
Drinking fountain	Offices, etc.	$^3/_8$" valve	0.25	—	0.25
Kitchen sink	Private	Faucet	1.0	1.0	1.4
Kitchen sink	Hotel, restaurant	Faucet	3.0	3.0	4.0
Laundry trays (1 to 3)	Private	Faucet	1.0	1.0	1.4
Lavatory	Private	Faucet	0.5	0.5	0.7
Lavatory	Public	Faucet	1.5	1.5	2.0
Service sink	Offices, etc.	Faucet	2.25	2.25	3.0
Shower head	Public	Mixing valve	3.0	3.0	4.0
Shower head	Private	Mixing valve	1.0	1.0	1.4
Urinal	Public	1" flush valve	10.0	—	10.0
Urinal	Public	$^3/_4$" flush valve	5.0	—	5.0
Urinal	Public	Flush tank	3.0	—	3.0
Washing machine (8 lb)	Private	Automatic	1.0	1.0	1.4
Washing machine (8 lb)	Public	Automatic	2.25	2.25	3.0
Washing machine (15 lb)	Public	Automatic	3.0	3.0	4.0
Water closet	Private	Flush valve	6.0	—	6.0
Water closet	Private	Flush valve	2.2	—	2.2
Water closet	Public	Flush valve	10.0	—	10.0
Water closet	Public	Flush valve	5.0	—	5.0
Water closet	Public or private	Flushometer tank	2.0	—	2.0

For SI: 1 inch = 25.4 mm, 1 pound = 0.454 kg.

a. For fixtures not listed , loads should be assumed by comparing the fixture to one listed using water in similar quantities and at similar rates. The assigned loads for fixtures with both hot and cold water supplies are given for separate hot and cold water loads and for total load. The separate hot and cold water loads being three-fourths of the total load for the fixture in each case.

TABLE E103.3(3)
TABLE FOR ESTIMATING DEMAND

SUPPLY SYSTEMS PREDOMINANTLY FOR FLUSH TANKS			SUPPLY SYSTEMS PREDOMINANTLY FOR FLUSH VALVES		
Load	Demand		Load	Demand	
(Water supply Fixture units)	(Gallons per minute)	(Cubic feet per minute)	(Water supply fixture units)	(Gallons per minute)	(Cubic feet per minute)
1	3.0	0.04104	—	—	—
2	5.0	0.0684	—	—	—
3	6.5	0.86892	—	—	—
4	8.0	1.06944	—	—	—
5	9.4	1.256592	5	15.0	2.0052
6	10.7	1.430376	6	17.4	2.326032
7	11.8	1.577424	7	19.8	2.646364
8	12.8	1.711104	8	22.2	2.967696
9	13.7	1.831416	9	24.6	3.288528
10	14.6	1.951728	10	27.0	3.60936
11	15.4	2.058672	11	27.8	3.716304
12	16.0	2.13888	12	28.6	3.823248
13	16.5	2.20572	13	29.4	3.930192
14	17.0	2.27256	14	30.2	4.037136
15	17.5	2.3394	15	31.0	4.14408
16	18.0	2.90624	16	31.8	4.241024
17	18.4	2.459712	17	32.6	4.357968
18	18.8	2.513184	18	33.4	4.464912
19	19.2	2.566656	19	34.2	4.571856
20	19.6	2.620128	20	35.0	4.6788
25	21.5	2.87412	25	38.0	5.07984
30	23.3	3.114744	30	42.0	5.61356
35	24.9	3.328632	35	44.0	5.88192
40	26.3	3.515784	40	46.0	6.14928
45	27.7	3.702936	45	48.0	6.41664
50	29.1	3.890088	50	50.0	6.684
60	32.0	4.27776	60	54.0	7.21872
70	35.0	4.6788	70	58.0	7.75344
80	38.0	5.07984	80	61.2	8.181216
90	41.0	5.48088	90	64.3	8.595624
100	43.5	5.81508	100	67.5	9.0234
120	48.0	6.41664	120	73.0	9.75864
140	52.5	7.0182	140	77.0	10.29336
160	57.0	7.61976	160	81.0	10.82808
180	61.0	8.15448	180	85.5	11.42964
200	65.0	8.6892	200	90.0	12.0312
225	70.0	9.3576	225	95.5	12.76644
250	75.0	10.026	250	101.0	13.50168
275	80.0	10.6944	275	104.5	13.96956
300	85.0	11.3628	300	108.0	14.43744
400	105.0	14.0364	400	127.0	16.97736
500	124.0	16.57632	500	143.0	19.11624
750	170.0	22.7256	750	177.0	23.66136
1,000	208.0	27.80544	1,000	208.0	27.80544
1,250	239.0	31.94952	1,250	239.0	31.94952
1,500	269.0	35.95992	1,500	269.0	35.95992
1,750	297.0	39.70296	1,750	297.0	39.70296
2,000	325.0	43.446	2,000	325.0	43.446
2,500	380.0	50.7984	2,500	380.0	50.7984
3,000	433.0	57.88344	3,000	433.0	57.88344
4,000	535.0	70.182	4,000	525.0	70.182
5,000	593.0	79.27224	5,000	593.0	79.27224

TABLE E103.3(4)
LOSS OF PRESSURE THROUGH TAPS AND TEES IN POUNDS PER SQUARE INCH (psi)

GALLONS PER MINUTE	SIZE OF TAP OR TEE (inches)						
	5/8	3/4	1	1 1/4	1 1/2	2	3
10	1.35	0.64	0.18	0.08	—	—	—
20	5.38	2.54	0.77	0.31	0.14	—	—
30	12.10	5.72	1.62	0.69	0.33	0.10	—
40	—	10.20	3.07	1.23	0.58	0.18	—
50	—	15.90	4.49	1.92	0.91	0.28	—
60	—	—	6.46	2.76	1.31	0.40	—
70	—	—	8.79	3.76	1.78	0.55	0.10
80	—	—	11.50	4.90	2.32	0.72	0.13
90	—	—	14.50	6.21	2.94	0.91	0.16
100	—	—	17.94	7.67	3.63	1.12	0.21
120	—	—	25.80	11.00	5.23	1.61	0.30
140	—	—	35.20	15.00	7.12	2.20	0.41
150	—	—	—	17.20	8.16	2.52	0.47
160	—	—	—	19.60	9.30	2.92	0.54
180	—	—	—	24.80	11.80	3.62	0.68
200	—	—	—	30.70	14.50	4.48	0.84
225	—	—	—	38.80	18.40	5.60	1.06
250	—	—	—	47.90	22.70	7.00	1.31
275	—	—	—	—	27.40	7.70	1.59
300	—	—	—	—	32.60	10.10	1.88

For SI: 1 inch = 25.4 mm, 1 pound per square inch - 6.895 kpa, 1 gallon per minute = 3.785 L/m.

TABLE E103.3(5)
ALLOWANCE IN EQUIVALENT LENGTHS OF PIPE FOR FRICTION LOSS IN VALVES AND THREADED FITTINGS (feet)

FITTING OR VALVE	PIPE SIZE (inches)							
	1/2	3/4	1	1 1/4	1 1/2	2	2 1/2	3
45-degree elbow	1.2	1.5	1.8	2.4	3.0	4.0	5.0	6.0
90-degree elbow	2.0	2.5	3.0	4.0	5.0	7.0	8.0	10.0
Tee, run	0.6	0.8	0.9	1.2	1.5	2.0	2.5	3.0
Tee, branch	3.0	4.0	5.0	6.0	7.0	10.0	12.0	15.0
Gate valve	0.4	0.5	0.6	0.8	1.0	1.3	1.6	2.0
Balancing valve	0.8	1.1	1.5	1.9	2.2	3.0	3.7	4.5
Plug-type cock	0.8	1.1	1.5	1.9	2.2	3.0	3.7	4.5
Check valve, swing	5.6	8.4	11.2	14.0	16.8	22.4	28.0	33.6
Globe valve	15.0	20.0	25.0	35.0	45.0	55.0	65.0	80.0
Angle valve	8.0	12.0	15.0	18.0	22.0	28.0	34.0	40.0

For SI: 1 inch = 25.4 mm, 1 foot = 304.8 mm, 1 degree = 0.0175 rad.

TABLE E103.3(6)
PRESSURE LOSS IN FITTINGS AND VALVES EXPRESSED AS EQUIVALENT LENGTH OF TUBE[a] (feet)

| NOMINAL OR STANDARD SIZE (inches) | FITTINGS | | | | Coupling | VALVES | | | |
| | Standard Ell | | 90-Degree Tee | | | Ball | Gate | Butterfly | Check |
	90 Degree	45 Degree	Side Branch	Straight Run					
$^3/_8$	0.5	—	1.5	—	—	—	—	—	1.5
$^1/_2$	1	0.5	2	—	—	—	—	—	2
$^5/_8$	1.5	0.5	2	—	—	—	—	—	2.5
$^3/_4$	2	0.5	3	—	—	—	—	—	3
1	2.5	1	4.5	—	—	0.5	—	—	4.5
$1^1/_4$	3	1	5.5	0.5	0.5	0.5	—	—	5.5
$1^1/_2$	4	1.5	7	0.5	0.5	0.5	—	—	6.5
2	5.5	2	9	0.5	0.5	0.5	0.5	7.5	9
$2^1/_2$	7	2.5	12	0.5	0.5	—	1	10	11.5
3	9	3.5	15	1	1	—	1.5	15.5	14.5
$3^1/_2$	9	3.5	14	1	1	—	2	—	12.5
4	12.5	5	21	1	1	—	2	16	18.5
5	16	6	27	1.5	1.5	—	3	11.5	23.5
6	19	7	34	2	2	—	3.5	13.5	26.5
8	29	11	50	3	3	—	5	12.5	39

For SI: 1 inch = 25.4 mm, 1 foot = 304.8 mm, 1 degree = 0.01745 rad.

a. Allowances are for streamlined soldered fittings and recessed threaded fittings. For threaded fittings, double the allowances shown in the table. The equivalent lengths presented above are based on a C factor of 150 in the Hazen-Williams friction loss formula. The lengths shown are rounded to the nearest half-foot.

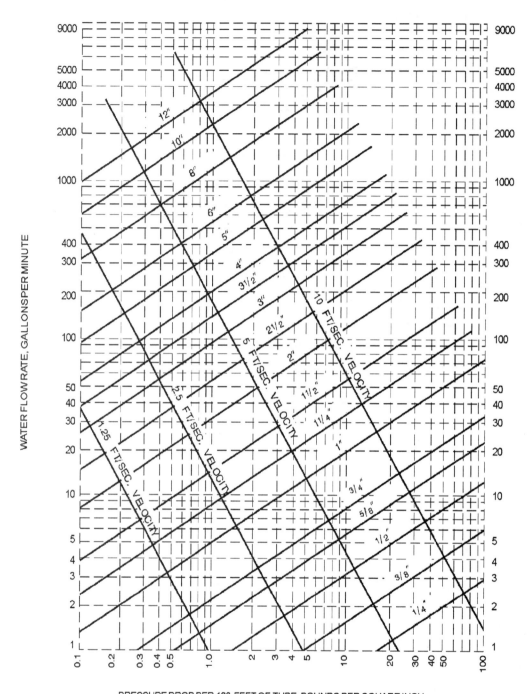

PRESSURE DROP PER 100 FEET OF TUBE, POUNDS PER SQUARE INCH

Note: Fluid velocities in excess of 5 to 8 feet/second are not usually recommended.

FIGURE E103.3(2)
FRICTION LOSS IN SMOOTH PIPE[a]
(TYPE K, ASTM B 88 COPPER TUBING)

For SI: 1 inch - 25.4 mm, 1 foot = 304.8 mm, 1 gpm = 3.785 L/m, 1 psi = 6.895 kPa,
1 foot per second = 0.305 m/s.

a. This chart applies to smooth new copper tubing with recessed (Streamline) soldered joints and
to the actual sizes of types indicated on the diagram.

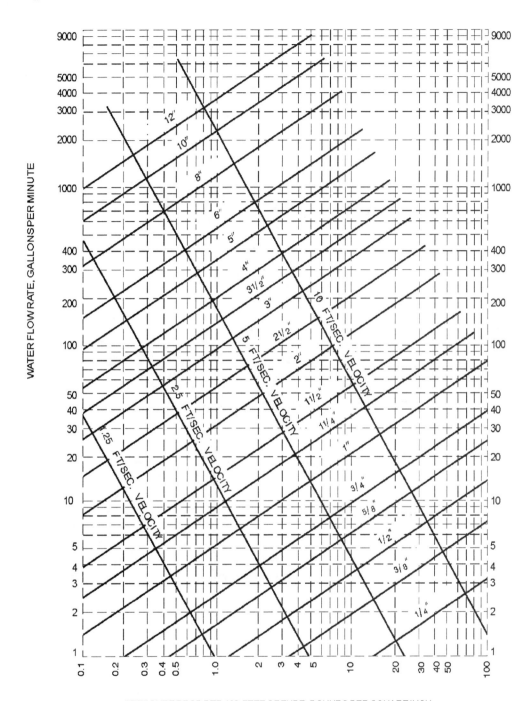

PRESSURE DROP PER 100 FEET OF TUBE, POUNDS PER SQUARE INCH

Note: Fluid velocities in excess of 5 to 8 feet/second are not usually recommended.

FIGURE E103.3(3)
FRICTION LOSS IN SMOOTH PIPE[a]
(TYPE L, ASTM B 88 COPPER TUBING)

For SI: 1 inch - 25.4 mm, 1 foot = 304.8 mm, 1 gpm = 3.785 L/m, 1 psi = 6.895 kPa,
1 foot per second = 0.305 m/s.

a. This chart applies to smooth new copper tubing with recessed (Streamline) soldered
joints and to the actual sizes of types indicated on the diagram.

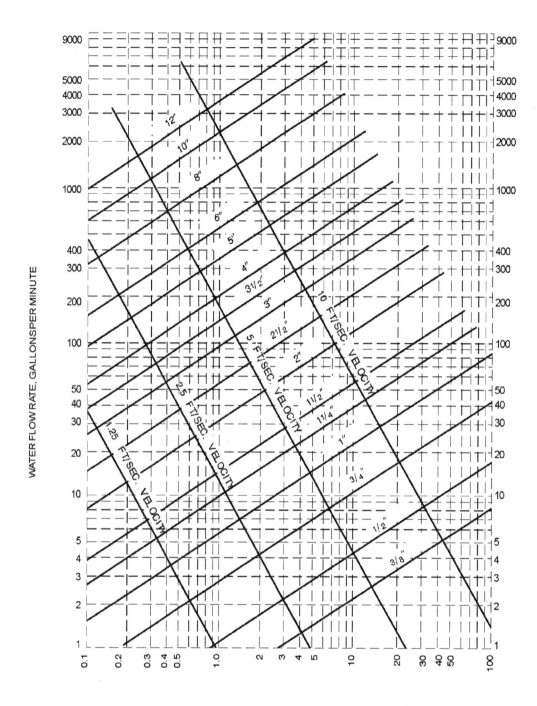

PRESSURE DROP PER 100 FEET OF TUBE, POUNDS PER SQUARE INCH

Note: Fluid velocities in excess of 5 to 8 feet/second are not usually recommended.

FIGURE E103.3(4)
FRICTION LOSS IN SMOOTH PIPE[a]
(TYPE M, ASTM B 88 COPPER TUBING)

For SI: 1 inch - 25.4 mm, 1 foot = 304.8 mm, 1 gpm = 3.785 L/m, 1 psi = 6.895 kPa,
1 foot per second = 0.305 m/s.

a. This chart applies to smooth new copper tubing with recessed (Streamline) soldered
joints and to the actual sizes of types indicated on the diagram.

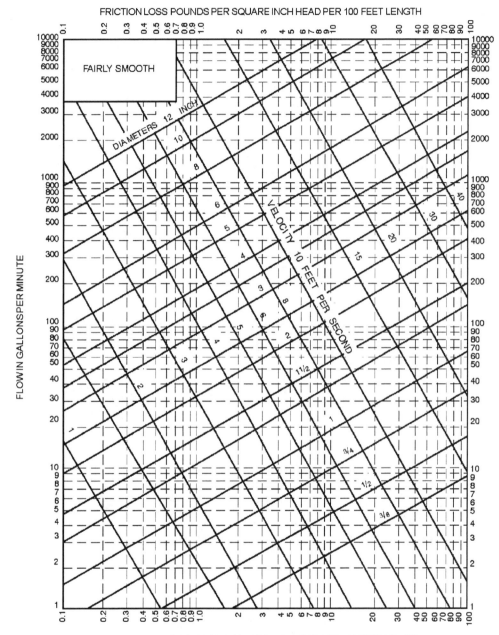

FIGURE E103.3(5)
FRICTION LOSS IN FAIRLY SMOOTH PIPE[a]

For SI: 1 inch - 25.4 mm, 1 foot = 304.8 mm, 1 gpm = 3.785 L/m, 1 psi = 6.895 kPa, 1 foot per second = 0.305 m/s.

a. This chart applies to smooth new steel (fairly smooth) pipe and to actual diameters of standard-weight pipe.

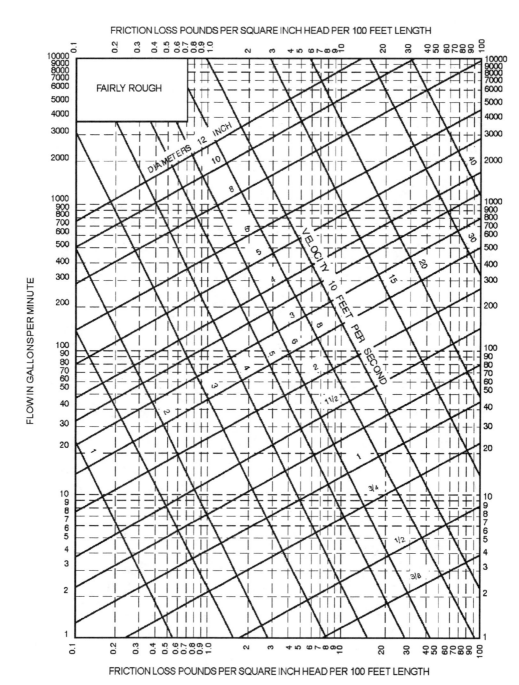

FRICTION LOSS POUNDS PER SQUARE INCH HEAD PER 100 FEET LENGTH

FIGURE E103.3(6)
FRICTION LOSS IN FAIRLY ROUGH PIPE[a]

For SI: 1 inch - 25.4 mm, 1 foot = 304.8 mm, 1 gpm = 3.785 L/m, 1 psi = 6.895 kPa,
1 foot per second = 0.305 m/s.

a. This chart applies to fairly rough pipe and to actual diameters which in general will
be less than the actual diameters of the new pipe of the same kind.

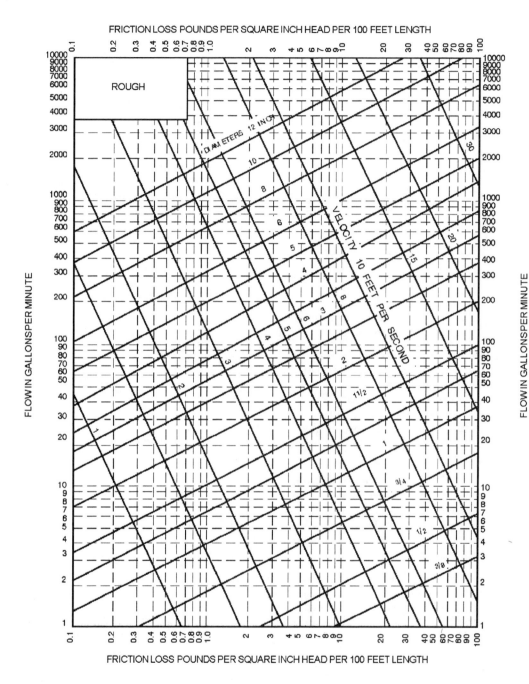

FRICTION LOSS POUNDS PER SQUARE INCH HEAD PER 100 FEET LENGTH

FRICTION LOSS POUNDS PER SQUARE INCH HEAD PER 100 FEET LENGTH

FIGURE E103.3(7)
FRICTION LOSS IN FAIRLY ROUGH PIPE[a]

For SI: 1 inch - 25.4 mm, 1 foot = 304.8 mm, 1 gpm = 3.785 L/m, 1 psi = 6.895 kPa,
1 foot per second = 0.305 m/s.

a. This chart applies to very rough pipe and existing pipe and to their actual diameters.

SECTION E201
SELECTION OF PIPE SIZE

E 201.1 Size of water-service mains, branch mains and risers. The minimum size water service pipe shall be $^3/_4$ inch (19.1 mm). The size of water service mains, branch mains and risers shall be determined according to water supply demand [gpm (L/m)], available water pressure [psi (kPa)] and friction loss due to the water meter and developed length of pipe [feet (m)], including equivalent length of fittings. The size of each water distribution system shall be determined according to the procedure outlined in this section or by other design methods conforming to acceptable engineering practice and approved by the code official:

1. Supply load in the building water-distribution system shall be determined by total load on the pipe being sized, in terms of water-supply fixture units (w.s.f.u.), as shown in Table E103.3(2). For fixtures not listed, choose a w.s.f.u. value of a fixture with similar flow characteristics.

2. Obtain the minimum daily static service pressure [psi (kPa)] available (as determined by the local water authority) at the water meter or other source of supply at the installation location. Adjust this minimum daily static pressure [psi (kPa)] for the following conditions:

   2.1. Determine the difference in elevation between the source of supply and the highest water supply outlet. Where the highest water supply outlet is located above the source of supply, deduct 0.5 psi (3.4 kPa) for each foot (0.3 m) of difference in elevation. Where the highest water supply outlet is located below the source of supply, add 0.5 psi (3.4 kPa) for each foot (0.3 m) of difference in elevation.

   2.2. Where a water pressure reducing valve is installed in the water distribution system, the minimum daily static water pressure available is 80 percent of the minimum daily static water pressure at the source of supply or the set pressure downstream of the pressure reducing valve, whichever is smaller.

   2.3. Deduct all pressure losses due to special equipment such as a backflow preventer, water filter and water softener. Pressure loss data for each piece of equipment shall be obtained through the manufacturer of such devices.

   2.4. Deduct the pressure in excess of 8 psi (55 kPa) due to installation of the special plumbing fixture, such as temperature controlled shower and flushometer tank water closet.

   Using the resulting minimum available pressure, find the corresponding pressure range in Table E201.1.

3. The maximum developed length for water piping is the actual length of pipe between the source of supply and the most remote fixture, including either hot (through the water heater) or cold water branches multiplied by a factor of 1.2 to compensate for pressure loss through fittings.

   Select the appropriate column in Table E201.1 equal to or greater than the calculated maximum developed length.

4. To determine the size of water service pipe, meter and main distribution pipe to the building using the appropriate table, follow down the selected "maximum developed length" column to a fixture unit equal to, or greater than the total installation demand calculated by using the "combined" water supply fixture unit column of Table E201.1. Read the water service pipe and meter sizes in the first left-hand column and the main distribution pipe to the building in the second left-hand column on the same row.

5. To determine the size of each water distribution pipe, start at the most remote outlet on each branch (either hot or cold branch) and, working back toward the main distribution pipe to the building, add up the water supply fixture unit demand passing through each segment of the distribution system using the related hot or cold column of Table E201.1. Knowing demand, the size of each segment shall be read from the second left-hand column of the same table and maximum developed length column selected in Steps 1 and 2, under the same or next smaller size meter row. In no case does the size of any branch or main need to be larger that the size of the main distribution pipe to the building established in Step 4.

TABLE E201.1
MINIMUM SIZE OF WATER METERS, MAINS AND DISTRIBUTION PIPING
BASED ON WATER SUPPLY FIXTURE UNIT VALUES (w.s.f.u.)

METER AND SERVICE PIPE (inches)	DISTRIBUTION PIPE (inches)	MAXIMUM DEVELOPMENT LENGTH (feet)									
Pressure Range 30 to 39 psi		40	60	80	100	150	200	250	300	400	500
$^3/_4$	$^1/_2$ a	2.5	2	1.5	1.5	1	1	0.5	0.5	0	0
$^3/_4$	$^3/_4$	9.5	7.5	6	5.5	4	3.5	3	2.5	2	1.5
$^3/_4$	1	32	25	20	16.5	11	9	7.8	6.5	5.5	4.5
1	1	32	32	27	21	13.5	10	8	7	5.5	5
$^3/_4$	$1^{-1}/_4$	32	32	32	32	30	24	20	17	13	10.5
1	$1^{-1}/_4$	80	80	70	61	45	34	27	22	16	12
$1^{-1}/_2$	$1^{-1}/_4$	80	80	80	75	54	40	31	25	17.5	13
1	$1^{-1}/_2$	87	87	87	87	84	73	64	56	45	36
$1^{-1}/_2$	$1^{-1}/_2$	151	151	151	151	117	92	79	69	54	43
2	$1^{-1}/_2$	151	151	151	151	128	99	83	72	56	45
1	2	87	87	87	87	87	87	87	87	87	86
$1^{-1}/_2$	2	275	275	275	275	258	223	196	174	144	122
2	2	365	365	365	365	318	266	229	201	160	134
2	$2^{-1}/_2$	533	533	533	533	533	495	448	409	353	311

METER AND SERVICE PIPE (inches)	DISTRIBUTION PIPE (inches)	MAXIMUM DEVELOPMENT LENGTH (feet)									
Pressure Range 40 to 49 psi		40	60	80	100	150	200	250	300	400	500
$^3/_4$	$^1/_2$ a	3	2.5	2	1.5	1.5	1	1	0.5	0.5	0.5
$^3/_4$	$^3/_4$	9.5	9.5	8.5	7	5.5	4.5	3.5	3	2.5	2
$^3/_4$	1	32	32	32	26	18	13.5	10.5	9	7.5	6
1	1	32	32	32	32	21	15	11.5	9.5	7.5	6.5
$^3/_4$	$1^{-1}/_4$	32	32	32	32	32	32	32	27	21	16.5
1	$1^{-1}/_4$	80	80	80	80	65	52	42	35	26	20
$1^{-1}/_2$	$1^{-1}/_4$	80	80	80	80	75	59	48	39	28	21
1	$1^{-1}/_2$	87	87	87	87	87	87	87	78	65	55
$1^{-1}/_2$	$1^{-1}/_2$	151	151	151	151	151	130	109	93	75	63
2	$1^{-1}/_2$	151	151	151	151	151	139	115	98	77	64
1	2	87	87	87	87	87	87	87	87	87	87
$1^{-1}/_2$	2	275	275	275	275	275	275	264	238	198	169
2	2	365	365	365	365	365	349	304	270	220	185
2	$2^{-1}/_2$	533	533	533	533	533	533	533	528	456	403

(continued)

TABLE E201.1—continued
MINIMUM SIZE OF WATER METERS, MAINS AND DISTRIBUTION PIPING
BASED ON WATER SUPPLY FIXTURE UNIT VALUES (w.s.f.u.)

METER AND SERVICE PIPE (inches)	DISTRIBUTION PIPE (inches)	MAXIMUM DEVELOPMENT LENGTH (feet)									
Pressure Range 50 to 60 psi		40	60	80	100	150	200	250	300	400	500
$3/4$	$1/2$ [a]	3	3	2.5	2	1.5	1	1	1	0.5	0.5
$3/4$	$3/4$	9.5	9.5	9.5	8.5	6.5	0.5	4.5	4	3	2.5
$3/4$	1	32	32	32	32	25	18.5	14.5	12	9.5	8
1	1	32	32	32	32	30	22	16.5	13	10	8
$3/4$	$1\text{-}1/4$	32	32	32	32	32	32	32	32	29	24
1	$1\text{-}1/4$	80	80	80	80	80	68	57	48	35	28
$1\text{-}1/2$	$1\text{-}1/4$	80	80	80	80	80	75	63	53	39	29
1	$1\text{-}1/2$	87	87	87	87	87	87	87	87	82	70
$1\text{-}1/2$	$1\text{-}1/2$	151	151	151	151	151	151	139	120	94	79
2	$1\text{-}1/2$	151	151	151	151	151	151	146	126	97	81
1	2	87	87	87	87	87	87	87	87	87	87
$1\text{-}1/2$	2	275	275	275	275	275	275	275	275	247	213
2	2	365	365	365	365	365	365	365	329	272	232
2	$2\text{-}1/2$	533	533	533	533	533	533	533	533	353	486

METER AND SERVICE PIPE (inches)	DISTRIBUTION PIPE (inches)	MAXIMUM DEVELOPMENT LENGTH (feet)									
Pressure Range Over 60		40	60	80	100	150	200	250	300	400	500
$3/4$	$1/2$ [a]	3	3	3	2.5	2	1.5	1.5	1	1	0.5
$3/4$	$3/4$	9.5	9.5	9.5	9.5	7.5	6	5	4.5	3.5	3
$3/4$	1	32	32	32	32	32	24	19.5	15.5	11.5	9.5
1	1	32	32	32	32	32	28	28	17	12	9.5
$3/4$	$1\text{-}1/4$	32	32	32	32	32	32	32	32	32	30
1	$1\text{-}1/4$	80	80	80	80	80	80	69	60	46	36
$1\text{-}1/2$	$1\text{-}1/4$	80	80	80	80	80	80	76	65	50	38
1	$1\text{-}1/2$	87	87	87	87	87	87	87	87	87	84
$1\text{-}1/2$	$1\text{-}1/2$	151	151	151	151	151	151	151	144	114	94
2	$1\text{-}1/2$	151	151	151	151	151	151	151	151	118	97
1	2	87	87	87	87	87	87	87	87	87	87
$1\text{-}1/2$	2	275	275	275	275	275	275	275	275	275	252
2	2	365	368	368	368	368	368	368	368	318	273
2	$2\text{-}1/2$	533	533	533	533	533	533	533	533	533	533

a. Minimum size for building supply is $3/4$-inch pipe.

APPENDIX F
STRUCTURAL SAFETY

SECTION F101
CUTTING, NOTCHING AND
BORING IN WOOD MEMBERS

F101.1 Joist notching. Notches on the ends of joists shall not exceed one-fourth the joist depth. Holes bored in joists shall not be within 2 inches (51 mm) of the top or bottom of the joist, and the diameter of any such hole shall not exceed one-third the depth of the joist. Notches in the top or bottom of joists shall not exceed one sixth the depth and shall not be located in the middle third of the span.

F101.2 Stud cutting and notching. In exterior walls and bear-ing partitions, any wood stud is permitted to be cut or notched to a depth not exceeding 25 percent of its width. Cutting or notching of studs to a depth not greater than 40 percent of the width of the stud is permitted in nonbearing partitions supporting no loads other than the weight of the partition.

F101.3 Bored holes. A hole not greater in diameter than 40 per-cent of the stud width is permitted to be bored in any wood stud. Bored holes not greater than 60 percent of the width of the stud is permitted in nonbearing partitions or in any wall where each bored stud is doubled, provided not more than two such successive doubled studs are so bored. In no case shall the edge of the bored hole be nearer than 0.625 inch (15.9 mm) to the edge of the stud. Bored holes shall not be located at the same section of stud as a cut of notch.

F101.4 Cutting, notching and boring holes in structural steel framing. The cutting, notching and boring of holes in structural steel framing members shall be as prescribed by the registered design professional.

F101.5 Cutting, notching and boring holes in cold-formed steel framing. Flanges and lips of load-bearing cold-formed steel framing members shall not be cut or notched. Holes in webs of load-bearing cold-formed steel framing members shall be permitted along the centerline of the web of the framing member and shall not exceed the dimensional limitations, penetration spacing or minimum hole edge distance as prescribed by the registered design professional. Cutting, notching and boring holes of steel floor/roof decking shall be as prescribed by the registered design professional.

F101.6 Cutting, notching and boring holes in nonstructural cold-formed steel wall framing. Flanges and lips of nonstructural cold-formed steel wall studs shall not be cut or notched. Holes in webs of nonstructural cold-formed steel wall studs shall be permitted along the centerline of the web of the framing member, shall not exceed 1.5 inches (38 mm) in width or 4 inches (102 mm) in length, and the holes shall not be spaced less than 24 inches (610 mm) center to center from another hole or less than 10 inches (254 mm) from the bearing end.

APPENDIX G

VACUUM DRAINAGE SYSTEM

SECTION G101
VACUUM DRAINAGE SYSTEM

G101.1 Scope. This appendix provides general guidelines for the requirements for vacuum drainage systems.

G101.2 General requirements.

G101.2.1 System design. Vacuum drainage systems shall be designed in accordance with manufacturer's recommendations. The system layout, including piping layout, tank assemblies, vacuum pump assembly and other components/designs necessary for proper function of the system shall be per manufacturer's recommendations. Plans, specifications and other data for such systems shall be submitted to the local administrative authority for review and approval prior to installation.

G101.2.2 Fixtures. Gravity-type fixtures used in vacuum drainage systems shall comply with Chapter 4 of this code.

G101.2.3 Drainage fixture units. Fixture units for gravity drainage systems which discharge into or receive discharge from vacuum drainage systems shall be based on values in Chapter 7 of this code.

G101.2.4 Water supply fixture units. Water supply fixture units shall be based on values in Chapter 6 of this code with the addition that the fixture unit of a vacuum-type water closet shall be "1."

G101.2.5 Traps and cleanouts. Gravity-type fixtures shall be provided with traps and cleanouts in accordance with Chapter 10 of this code.

G101.2.6 Materials. Vacuum drainage pipe, fitting and valve materials shall be as recommended by the vacuum drainage system manufacturer and as permitted by this code.

G101.3 Testing and demonstrations. After completion of the entire system installation, the system shall be subjected to a vacuum test of 19 inches (483 mm) of mercury and shall be operated to function as required by the administrative authority and the manufacturer. Recorded proof of all tests shall be submitted to the administrative authority.

G101.4 Written instructions. Written instructions for the operations, maintenance, safety and emergency procedures shall be provided by the building owner as verified by the administrative authority.

INDEX

EDITORIAL CHANGES – SECOND PRINTING

Page 34, 504.6.2: line 2 now reads . . . those materials listed in Section 605.4 or shall be tested, rated ...

Page 45, 608.13.3: line 4 now reads . . . CAN/CSA-B64.3.

Page 95, ASTM: Standard reference number now reads . . . D 2665—01

Page 98, NSF: Standard reference number now reads . . . 1996a